BUTTERFLIES AND MOTHS

A GUIDE TO THE MORE COMMON AMERICAN SPECIES

by
ROBERT T. MITCHELL
and
HERBERT S. ZIM

Illustrated by
ANDRE DURENCEAU

 Golden Books® • NEW YORK

FOREWORD

This book presents an introduction to American butterflies and moths. So numerous are North American species that only about four per cent have been included, but these were selected to include the most common, widespread, important, or unusual kinds. Special attention has been given to immature forms and to range maps.

Andre Durenceau deserves our special thanks for his magnificent art, so painstakingly done. The technical assistance of William D. Field has also been invaluable. The authors are also gratefully indebted to other specialists formerly or currently of the Smithsonian Institution, especially H. W. Capps, J. F. Gates Clarke, Douglas Ferguson, Ronald Hodges, and E. L. Todd. Among numerous others who contributed are W. A. Anderson, T. L. Bissell, J. H. Fales, R. S. Simmons, Richard Smith, and several entomologists of the U.S. Forest Service.

This Revised Edition includes recent changes in scientific and common names and geographical distributions, and it stresses conservation. Robert Robbins of the United States National Museum gave valuable technical assistance in the section on butterflies. New artwork was done by Ray Skibinski.

R. T. M.
H. S. Z.

CONTENTS

INTRODUCING LEPIDOPTERA

Butterflies and moths are most numerous in the tropics, but temperate areas have a bountiful supply of many species. Like all insects, they have three main body regions (head, thorax, and abdomen), three pairs of jointed legs, and one pair of antennae. Most have two pairs of wings. A few are wingless.

Insects that possess certain basic structures in common are classified into large groups or orders. Butterflies and moths are members of the order Lepidoptera, derived from the Greek *lepidos* for scales and *ptera* for wings. Their scaled wings distinguish them as a group from all other insects. When butterflies and moths are handled, the scales rub off as colored powder. Under a microscope, the colors and forms of the scales are amazing.

Lepidoptera is the largest order of insects next to Coleoptera (beetles). Beetles are estimated at about 280,000 species; Lepidoptera at 120,000, with about 10,000 species in North America. Lepidoptera is usually divided into three suborders: first, *Jugatae,* with about 250 primitive species that somewhat resemble caddisflies; second, *Frenatae,* most moths; and, third, *Rhopalocera,* the butterflies and skippers.

The suborder Rhopalocera is divided into two superfamilies: *Papilionoidea,* which includes 19 families of butterflies, and *Hesperioidea,* two families of skippers. Butterflies and skippers are easy to distinguish by the shape and position of their antennae (pp. 19 and 74).

The suborder Frenatae includes about fifty families of North American moths. No single feature will enable one to tell a moth from a butterfly or skipper, but a frenulum (p. 81) on the hindwing of most moths extends to the forewing, holding the wings together. The presence and position of simple eyes (ocelli) and leg spines, the nature

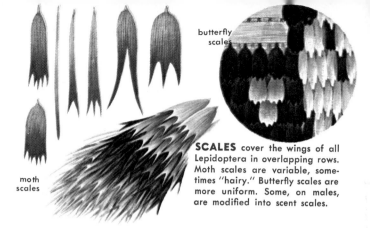

SCALES cover the wings of all Lepidoptera in overlapping rows. Moth scales are variable, sometimes "hairy." Butterfly scales are more uniform. Some, on males, are modified into scent scales.

butterfly scales

moth scales

of the antennae, and the shape and venation of the wings are used in moth identification. To make veins more visible for study, moisten the wings with alcohol.

This guide employs the common names of butterflies and moths for ease of use by beginners. But the book closely follows scientific classification of Lepidoptera. The 6 families of North American butterflies as herein named are those recognized in the collection of the U.S. National Museum. They are then broken down into genera (plural of genus), which in turn contain one or more species. To help you follow the organization, butterfly family names appear in red, butterfly genera and species names in black. Because they are so numerous, moths are dealt with mainly on the family and species levels.

Each species of Lepidoptera bears a double scientific name, such as *Pieris rapae* for the Cabbage Butterfly. *Pieris* is the name of the genus; *rapae* is the species name. See pp. 154-157 for scientific names of species illustrated in this book.

Io Moth

Polyphemus Moth

Eastern Tailed
Blue Butterfly

Sulphur
Butterfly

Comma
Butterfly

Monarch Butterfly

Tent Caterpillar
egg mass

EGGS OF BUTTERFLIES AND MOTHS

LIFE HISTORIES Lepidoptera develop by a complete metamorphosis, which is characterized by four distinct growth stages, as shown for the Gypsy Moth on p. 136. The egg hatches into a *larva*, or caterpillar, which grows and molts (sheds its skin) several times before transforming into a *pupa* from which a winged (usually) *adult* emerges later.

EGGS of Lepidoptera vary greatly in size and shape. Many are spherical but some kinds are flattened, conical, spindle- or barrel-shaped. Some eggs are smooth, but others are ornamented with ribs, pits, or grooves, or networks of fine ridges. Each egg has a small hole through which it is fertilized.

The adult female may lay eggs singly, in small clusters, or in one egg mass. Most often eggs are deposited on a plant that will serve as food for the larvae. Some eggs are laid on the ground, and the newly hatched larvae must seek their food plants. Eggs laid during the summer are usually thin-coated; those that overwinter before hatching have a thicker outer coat and are sometimes covered by "hair" from the moth. They may also be covered with a foamy layer, as shown for the Tent Caterpillar on p. 6.

Most eggs hatch in a few days. The larva, which can frequently be seen inside the egg just before hatching, eats its way out and sometimes also eats the eggshell.

LARVAE of Lepidoptera are caterpillars, though some are known as worms, slugs, or borers. North American caterpillars range in length from 0.2 inch to about 6 inches. Like the adult, the caterpillar has three body regions—head, thorax, and abdomen.

On each side of the head are tiny ocelli, or simple eyes, usually in a semi-circle, and a tiny antenna. The mouthparts include an upper lip (labrum), a pair of strong jaws (mandibles), two small sensory organs (palpi), and a lower lip (labium), which bears a pair of spinnerets, used for spinning silk threads.

On each of the three segments of the thorax is a pair of short jointed legs, ending in claws. On each side of the first thoracic segment is a spiracle, an opening for breathing.

MOTH LARVAE WITH STINGING SPINES OR HAIRS

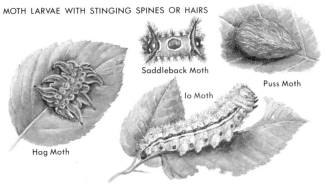

Saddleback Moth

Puss Moth

Io Moth

Hag Moth

The abdomen, usually composed of ten segments, bears two to five pairs of short, fleshy prolegs. Segment 10 bears the largest pair, the anal prolegs. Spiracles occur on each side of the first eight abdominal segments.

Most larvae feed actively throughout their lives. Some kinds mature in a few weeks, others in months. Some become dormant, or estivate, during the summer; others hibernate, overwintering in newly hatched, partly grown, or fully grown stages. Most kinds feed on leaves, but others feed on flowers, fruits, and seeds, or bore into stems and wood. A few species are scavengers and a small number prey on insects, especially plant lice. A few feed on animal products like wool, silk, or feathers.

As a larva grows, it sheds its skin, or molts, allowing for another growth period. Larvae in stages between molts are called instars. Early instars may differ from later ones in color, markings, and shape.

Caterpillars with horns and spines may appear treacherous, but only a few, such as the Io, Hag Moth, Puss Moth, Saddleback Caterpillar and related "slugs," have irritating spines or hairs to avoid.

PUPAE are the resting forms in which the larvae transform into adults. Most butterflies and moths in temperate regions spend the winter as pupae, though the pupal stage of some species lasts for only a few days or weeks. In a prepupal stage the larva loses its prolegs; later its mouthparts change from chewing mandibles to a long proboscis (if present in the adult), wings develop, and reproductive organs form. External factors such as temperature and moisture may trigger the changes, but the actual transformation is caused by hormones.

The butterfly larva, when mature, attaches itself to a firm support before changing to a naked pupa, known

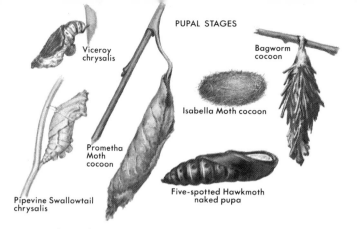

PUPAL STAGES

Viceroy chrysalis

Bagworm cocoon

Isabella Moth cocoon

Prometha Moth cocoon

Five-spotted Hawkmoth naked pupa

Pipevine Swallowtail chrysalis

as a chrysalis. Larvae of swallowtails, sulphurs, and whites deftly support theirs with a strong silk thread.

Most moth larvae, when full grown, burrow into the ground and pupate there in earthen cells. Others pupate amid dead leaves or debris on the ground, in hollow stems or decaying wood, sometimes with material drawn loosely together with silk. Hairy species usually mix their hairs with silk, making a flimsy cocoon. Silk Moth larvae spin tough papery silken cocoons that house their pupae. When emerging from these tight cocoons, moths secrete a fluid that softens the silk. Bagworm Moths construct cocoons around their bodies as they grow. At maturity they fasten the finished cocoons to twigs with silk.

While butterflies emerge easily from chrysalises, moths often exert great effort to break through cocoons or push their way up through the ground. Both emerge with soft small wings with miniature wing patterns. As fluids are pumped through the veins, the wings expand. Later the veins harden, providing a rigid support for the wing membrane.

9

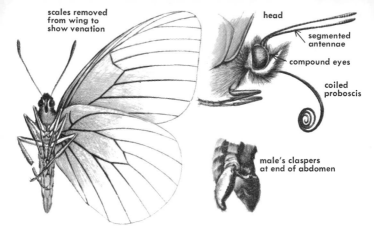

scales removed from wing to show venation

head

segmented antennae

compound eyes

coiled proboscis

male's claspers at end of abdomen

ADULT butterflies and moths have a pair of segmented antennae and a pair of large, rounded compound eyes on their heads. Many moths also have a pair of simple eyes. Butterflies and many moths have a coiled proboscis, which unrolls into a long sucking tube through which the adult feeds on nectar and other fluids. This tube may be as long as the adult's body.

Each of the three segments of the thorax bears a pair of five-jointed legs. Some groups of butterflies have the first pair of legs reduced, and females of the Bagworm Moth have no legs. A pair of membranous wings are attached to the 2nd and 3rd thoracic segments of most butterflies and moths but a few kinds are wingless. The vein pattern of wings is used in classification.

At the end of the ten-segmented abdomen are the sex organs. They are used in the accurate identification of many species. The female's abdomen is usually larger than the male's. The latter can be distinguished by the claspers of the sex organs which protrude as plate-like structures at the end of the last segment.

NATURAL ENEMIES of Lepidoptera abound. Various insects feed on them. So do spiders, birds, rodents, reptiles, amphibians, and night prowlers like skunks and raccoons. Parasitic insects lay eggs in and on caterpillars, eggs, or pupae, which then become food for the parasitic larvae. Bacteria, fungi, protozoa, and viruses cause diseases; unfavorable weather also takes its toll.

DEFENSES against such a host of destructive forces are necessary for survival. The capacity of females to lay hundreds of eggs is one. Camouflage, hiding from predators, is another. Other protective features are body markings that frighten enemies, and hairs, spines, or body juices unpleasant to them.

MAN is enemy no. 1. The destruction of favorable habitat from land development has led to a great decline in their numbers. Herbicides and pesticides kill them. Floodlights at malls, intersections, and athletic fields are lethal moth traps. Against man, they have no built-in defenses. For their survival, they are becoming more dependent on people who care, and so become involved in conservation.

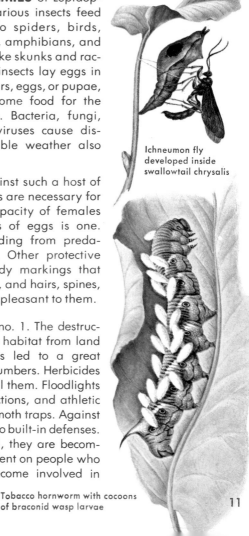

Ichneumon fly developed inside swallowtail chrysalis

Tobacco hornworm with cocoons of braconid wasp larvae

11

CONSERVATION is of growing importance. At least two species of butterflies are now extinct, and a number of other Lepidoptera have been listed as Threatened or Endangered. Here are some ways that you can help.

extinct Xerces Blue
(*Glaucopsyche xerces*)

JOIN a conservation-oriented organization, perhaps one of those listed below. Some 40 states now have Natural Heritage Programs that inventory their plant and animal life and make proposals for species of special concern. Contact your state office or Nature Conservancy to learn about local efforts where you can be helpful.

The Xerces Society, 10 Southwest Ash St., Portland, OR 97204. Dedicated to the preservation of arthropods and their habitats and promoting annual counts of butterflies in areas throughout the country.

The Lepidopterists' Society, c/o Julian P. Donahue, Asst. Sec., Natural History Museum of Los Angeles County, 900 Exposition Blvd., Los Angeles, CA 90007. An international society of specialists that publishes a journal of research papers and an annual summary of field observations of Lepidoptera of Canada and the U.S. as reported by members.

National Institute for Urban Wildlife, 10921 Trotting Ridge Way, Columbia, MD 21044. Focuses on conservation of urban and suburban areas.

The Nature Conservancy, 1800 North Kent St., Arlington, VA 22209. An outstanding conservator and manager of valuable habitats of rare and endangered plants and wildlife throughout the nation.

CREATE A BUTTERFLY GARDEN Plant such perennials as pussy willow, lilac, blueberry, *Clethra*, phlox, butterfly weed and butterfly bush, lantana, and such annuals as zinnia, French marigold, and single petunia in your garden to provide butterfly food throughout the season. Also plant appropriate food for larvae of the butterflies that come to feed as indicated in this book.

REAR butterflies and moths from eggs or larvae for release. Watch them grow and develop. See how they move, how they feed, what they do. Then return them to their preferred habitat.

Female moths confined in paper bags will often lay eggs there, but butterfly eggs are harder to obtain. Look for them when you see a butterfly exploring the leaves rather than the blossoms of a plant. Chewed or missing leaves on a plant are clues to the presence of caterpillars nearby that you might collect for rearing.

Eggs and small larvae at first can be kept in tightly sealed, clear polyethylene sandwich bags, together with a few leaves of their plant food. Keep each kind in a separate bag. Keep the bags out of the sun or excessive heat. Remove the larval droppings every day or so. Reverse the bag and add fresh leaves whenever the old ones start to yellow or to dry out. Use leaves of the same species of plant, and do not bag them when they are wet.

Transfer 2-inch larvae to larger clear bags or to tightly sealed cans, such as 1-lb coffee cans. To watch developments, the "bouquet" set-up can be used (see the illustration).

"Bouquet" set-up for rearing

Large plastic bag set-up for rearing many larvae of the same species

When a butterfly larva is almost full grown, put a stick in the can or bag to encourage the larva to form its chrysalis on it.

Large numbers of late-instar larvae of the same species and age can be reared in big freezer bags containing branches of the food plant, as illustrated. To clean out droppings, untie and allow them to fall through the opening.

Large numbers of larvae can be reared out-doors with less care by enclosing them in a strong net bag pulled over the end of a growing branch of a tree or bush and tied securely farther down the branch.

Caterpillars that make cocoons, such as Silk Moths and Tiger Moths, can be reared like those of butterflies. However, the larvae of Regal Moths and most noctuids must be given a few inches of damp (not wet) sterile soil or peat moss into which to burrow when full grown. The resulting pupae can be overwintered in sealed plastic sandwich bags (along with the damp medium) in a refrigerator. Keep overwintering cocoons and chrysalises outdoors, in cages to protect them from predators.

Adults emerging from pupae in the following seasons must be given ample room for spreading their wings and a rough surface for climbing to a perch. For chrysalises and cocoons only, a screened cage is needed. It can be made from a rolled section of wire screening or small-mesh hardware cloth; use paper plates for top and bottom.

A cylindrical cardboard rolled-oats box makes an ideal emergence cage for cocoons and chrysalises. When the open top is covered with a nylon stocking (held in place by tucking the leg and toe under a loop of material near the rim), the adult can be captured and brought to hand by extending the leg above the open top as the adult flies into the leg trying to escape.

Homemade cage

For pupa formed in soil, use topless round cans with a rough (rusty) surface for climbing, covered with gauze, netting, or a stocking. Then cover the can with a piece of clear polyethylene to keep the soil from drying out and to let you see any emerging moths.

After you have completed your observations, return the adults to their preferred habitat.

COLLECT SPARINGLY—and be sure to follow laws concerning endangered species. The chief aims of a collector should be to obtain subjects for rearing or for making a study collection. Usually, adults are collected while feeding at flowers or bait. They are rarely caught on the wing. The specimen is quickly transferred to a killing jar. Later it is mounted, spread, labeled, and cataloged. To make an acceptable study collection, some items must be purchased from a biological supply house (see below). Others can be homemade.

American Biological Supply Co., 1330 Dillon Heights Ave., Baltimore, MD 21228

BioQuip Products, P.O. Box 61, Santa Monica, CA 90406

Carolina Biological Supply Co., 2700 York Rd., Burlington, NC 27215

Ward's Natural Science Establishment, Inc., 5100 West Henrietta Rd., Rochester, NY 14692-9012

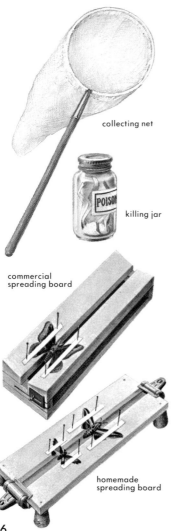

COLLECTING NETS should be lightweight, with rim 12 to 15 inches in diameter. Strong nylon net bag should be 27 to 32 inches deep, roughly funnel-shaped but not sharply pointed at the end.

collecting net

KILLING JARS should have wide mouths and seal tightly. Put enough paper toweling in the bottom to absorb a teaspoon to a tablespoon of liquid. To use, add enough ethyl acetate or carbon tetrachloride to saturate the paper; pour off any excess. Specimens too stiff for mounting can be relaxed by enclosing for a few hours in a plastic food storage box on a sheet of plastic spread over water-saturated paper toweling.

killing jar

commercial spreading board

SPREADING BOARDS are made of soft wood with a center channel in which the body of the specimen fits. When specimen is relaxed, insert insect pin straight down through center of thorax, ¼-inch from head; then stick pin into center of channel until wings are level with upper surface of board. If necessary, brace by inserting a pin in the channel on each side of specimen's body at base of hindwings. Spread wings gently with foreceps and pins so edges of forewings are at right angles to body and hindwings are in a normal position. Pin wings in place with paper strips. Neatly position antennae. Allow several days for drying.

homemade spreading board

INSECT PINS are made of special rust-resistant steel and come in several sizes. Size 3 can be used for all but small butterflies and moths.

LABELS should be placed on the pin of each specimen when it is removed from the spreading board. They should tell at least where, when, and by whom the specimen was taken. Labels should be neat and small. Sheets of typewritten labels can be photographically reduced to make small but readable labels.

The species label, containing the scientific name of the specimen, is larger and is usually pinned to the bottom of the storage box by the specimen pin. Supplement your collection data with a notebook of observations and field records.

STORAGE AND DISPLAY BOXES are of several kinds. The Schmitt box, with cork bottom, glazed paper lining, and tight-fitting lid, is ideal for housing a study collection. Supply houses also have less expensive boxes. The beginner can get along with a tightly lidded box that has a ⅜-inch layer of polyethylene foam, soft composition board, or balsa wood fitted into the bottom to keep the pins secure. All boxes must be treated periodically with paradichlorobenzine crystals to keep out destructive insect pests. If the box will be stored horizontally, crystals can be scattered on the bottom. If box is

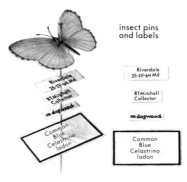

insect pins and labels

Riverdale
25-IV-64 Md

RTMitchell
Collector

on dogwood

Common
Blue
Celastrina
ladon

Schmitt box

stored on edge, specimens must be protected from the crystals. Put crystals in a small (closed) envelope inserted into a larger one (with flap removed); glue to lid of box. Pests found in a box can be killed by putting the box into a freezer for a few days.

OTHER STUDIES Anyone watching insects in the field may discover new facts about them. The behavior of some species is still little known, and a careful observer can make real contributions to our knowledge. Keep detailed and accurate records of your observations, being sure to provide answers to the questions: *What, where, when, how, how many,* and *how long.* Become skilled in close-up photography of the different stages of development of butterflies and moths, capturing their beauty and interesting behavior. Showing your pictures or slides to the young and old in your community will generate interest in these fascinating creatures among other people and promote their conservation.

BOOKS offer the quickest way to extend your knowledge about butterflies and moths. Try the following:

Covell, Charles V., Jr., *A Field Guide to the Moths of Eastern North America.* Boston: Houghton Mifflin Co., 1984.

Ehrlich, Paul R., and Anne H., *How to Know the Butterflies.* Dubuque, IA: Wm. C. Brown Co., 1961.

Ferguson, Douglas C., *Bombycoidea (Saturniidae, Silk and Regal Moths),* Fascicle 20, Parts 2A and 2B, in *The Moths of America North of Mexico* Series, edited by R. B. Dominick et al. London: E. W. Classey Ltd. & R. B. D. Publications Inc., 1971/72.

Hodges, R. W., *Sphingoidea* (Hawkmoths), Fascicle 21, in *The Moths of America North of Mexico* Series, edited by R. B. Dominick et al. London: E. W. Classey Ltd. & R. B. D. Publications Inc., 1971.

Holland, W. J., *The Moth Book.* Garden City, NY: Doubleday & Co., 1908. Reprinted New York: Dover Publications, 1968.

Howe, William H., *The Butterflies of North America.* Garden City, NY: Doubleday & Co., Inc., 1975.

Klots, A. B., *A Field Guide to the Butterflies of Eastern North America.* Boston : Houghton Mifflin Co., 1977.

Pyle, Robert M., *The Audubon Society Field Guide to North American Butterflies.* New York: Alfred A. Knopf, 1981.

Tekulsky, Mathew, *Butterfly Garden.* Boston, MA: Harvard Common Press, 1985.

BUTTERFLIES

Butterflies number about 700 species in North America north of Mexico. Recent research shows that some long considered distinct species are merely varieties (sub-species) of others. A few of these are included in this book.

Red-spotted Purple

Butterflies usually fly by day, and rest with their wings erect. Antennae of butterflies are clublike, ending in a swollen tip. Skippers (p. 74) have similar antennae that often turn back in a hook. Antennae of moths are seldom clublike, and are often feathery. Butterflies have a projection (the enlarged humeral lobe) on each hindwing that underlaps the front pair of wings and holds the wings together. Most butterflies pupate as an unprotected chrysalis which hangs freely from a plant or other support. Only a few butterflies are considered destructive, but many moths are pests.

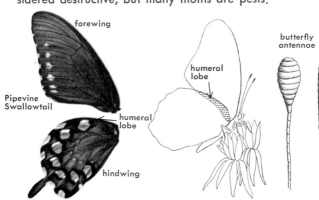

forewing

Pipevine
Swallowtail

humeral
lobe

hindwing

humeral
lobe

butterfly
antennae

SWALLOWTAILS are the largest and best known of our butterflies. They are found the world over, mainly in the tropics, and most are brightly colored. There are some two dozen species in North America, most having characteristic tail-like projections from the hindwings, usually one but also two or three in some species. These tails are lacking in the parnassius group (p. 29) and some others. In many species, the females look different from the males in size and markings.

Most swallowtails lay their spherical eggs singly on the food plants, mainly trees and shrubs. Later, the larvae may be found resting on a silken mat in a rolled leaf. Most species have an orange, fleshy, horn-like organ behind the head that emerges when the larva is disturbed and gives off a disagreeable odor.

When a full-grown swallowtail caterpillar has selected a place to form its chrysalis, it fastens its hindmost feet securely with silk and loops a tough silk thread behind its body, fastening the ends to the support as a kind of "safety belt." Soon the caterpillar sheds its skin, becoming a rough, angular chrysalis, and usually spends the winter in this form. Adults are often seen at flowers and are attracted to wet soil, puddles, or ponds.

PIPEVINE SWALLOWTAIL larva or caterpillar feeding on pipevine. Chrysalis to the right.

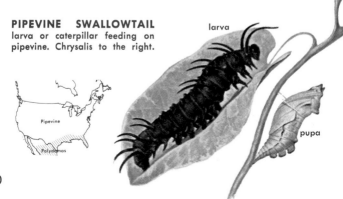

larva

pupa

Pipevine

Polydamas

PIPEVINE SWALLOWTAIL, is more common in the southern part of its range. The female has slightly larger whitish spots along the margin of the front wings and less colorful hind wings. She lays orange eggs in clusters on pipevine and snakeroot, on which the larvae feed.

3.0-4.5"

3.0-4.0"

POLYDAMAS SWALLOWTAIL is our only tailless black species. The caterpillar and chrysalis are similar to those of the Pipevine Swallowtail, but the caterpillar has reddish tentacles. It also feeds on pipevine. Note the greenish tint to the hindwings.

Black Swallowtail
2.8-3.5"

male

female

BLACK SWALLOWTAIL, also called the Common Eastern, or Parsnip, Swallowtail, has variable markings. Some males have hardly any blue on the hindwing. The spots may be larger or may be orange instead of yellow. Occasionally the two rows of spots on the forewing are fused into large triangular areas, or the spots may be greatly reduced. The hindwing in some forms is almost entirely yellow, tinged with orange.

The Black Swallowtail is found in open fields and woodland meadows. It frequents clover and flower gardens, always flying near the ground. The yellowish, ovoid eggs are laid on wild and cultivated plants of the carrot family, such as parsley, parsnip, celery, and carrot. When small, the larva, like that of most swallowtails, is dark brown with a white saddle mark. It becomes green, as illustrated, as it matures. There are two broods of Black Swallowtails annually in the North and at least three in the South.

BAIRD'S SWALLOWTAIL is very closely related to the Black Swallowtail and is also called the Western Black Swallowtail. The males are similar, but there is some difference in the females, which have less yellow than the eastern species. This butterfly is variable; some are quite yellow, some almost entirely black. The caterpillar feeds on sagebrush *(Artemesia),* and not on plants of the carrot family. There is one brood annually.

ANISE SWALLOWTAIL is probably the most common swallowtail west of the Rocky Mountains. The early stages of the larva closely resemble those of the Black Swallowtail in form and color. The larva feeds mainly on anise or Sweet Fennel of the carrot family. The adult female looks very much like the male.

Baird's Swallowtail
3.3-3.5"
female

Anise Swallowtail
2.5-3.5"

Black Swallowtail

pupa

larva

hindwings

Black
2.8-3.5"

Indra
2.0-3.0"

INDRA, or Mountain Swallowtail, can be distinguished from Baird's and Black by its short tail. Its light wing band is variable in size. Another species, the Short-tailed Swallowtail, occurs in eastern Canada. 2.0-3.0"

23

4.5-5.5"

PALAMEDES SWALLOWTAIL

PALAMEDES SWALLOW-TAIL rivals the Giant (p. 25) in size. Common in the South, it prefers the margins of swampy woods, where in slow flight it sometimes rises to the tops of tall trees. There are two or three broods a year. Eggs are laid on food plants—bay, magnolia, and sassafras. The caterpillar resembles that of the Spicebush (p. 26), but the reddish spot that appears on the third body segment is not so distinct and lacks the black ring.

ALASKAN SWALLOWTAIL

ALASKAN SWALLOWTAIL is a smaller and yellower variety of the European, or Old World, Swallowtail, common in Europe and Asia. Another variety of this Old World species is found near Hudson Bay in Canada. The caterpillar resembles the Black Swallowtail's (p. 22) and feeds on plants of the carrot family.

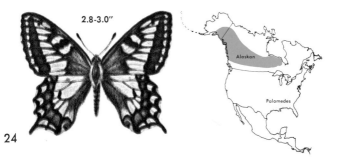

2.8-3.0"

Alaskan

Palamedes

4.0-5.5"

GIANT SWALLOWTAIL caterpillars are known as Orange Dogs or Orange Puppies in the South, where they do occasional damage to citrus trees, especially in young groves. Four or five hundred eggs may be laid by one female, deposited one at a time near tips of leaves or branches. The caterpillars feed on Prickly Ash and the Hop Tree in addition to citrus.

The Giant Swallowtail is more common in the southern part of its range, where it is likely to have three instead of two broods. First-brood adults emerge from chrysalids in May. The Giant Swallowtail has a leisurely flight, sometimes sailing with outstretched wings, which then show the bright yellow underside in contrast to the brown upperside. The Giant frequents open fields and gardens, sipping nectar from flowers and moisture from mud.

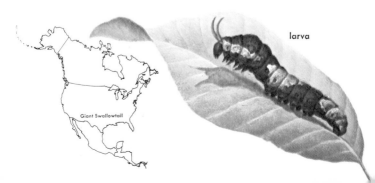

Giant Swallowtail

larva

4.0-5.0"

Hindwing of female is blue instead of greenish.

SPICEBUSH SWALLOWTAIL

is sometimes called the Green-clouded Swallowtail because the male's hindwing has a pronounced greenish tone. This species has a red-orange spot on the upper margin of the hindwing above.

These swallowtails frequent low, damp woods, visiting open fields less often than many other swallowtails. They are active, steady fliers and seldom alight. Numbers of them often gather at puddles on woodland roads or at other wet places. This butterfly has several geographic forms, some with larger yellow spots and other variations.

There are two broods in the North, three in the South. Butterflies of the first brood, which emerge from chrysalids in late April or early May, after spending the winter in the pupal stage, are smaller than those which emerge from later broods in the summer.

The larva feeds on Spicebush, Sassafras, Sweet Bay, and Prickly Ash. Like other swallowtail caterpillars it forms a mat of silk on the upper surface of a leaf; then it draws the leaf together and hides when not feeding. As it grows the larva constructs new and larger shelters until it is ready to pupate.

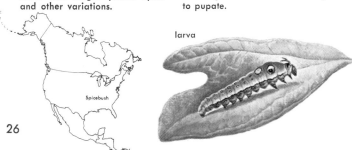

Spicebush

larva

Eastern Tiger Swallowtail
4.0-5.5"

male

female
light phase

female
dark phase

larva

EASTERN TIGER SWAL-LOWTAIL shows difference in color between sexes. Females are dimorphic (show two color forms); some are yellow and others dark brown. The dark form is uncommon in the North. The larva of the Eastern Tiger Swallowtail feeds mostly on Wild Cherry and Tuliptrees.

WESTERN TIGER SWAL-LOWTAIL is not dimorphic. It differs from the Eastern Swallowtail in having the spots on the underside of the forewing merge to form a band (see below). The caterpillar is like that of the Eastern Tiger Swallowtail but feeds on willow, poplar, and hops, plants of moist western areas.

Underside of forewing:

Eastern Tiger
spotted

Western Tiger
banded

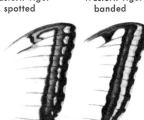

4.0-6.0"

TWO-TAILED SWALLOW-TAIL, our largest butterfly, occurs from British Columbia to Calif. and eastward to western Texas and Montana. The cater-pillar feeds on cherry, hoptree, ash, privet, and shadbush. Probably breeds twice each year. The smaller Three-tailed Swallowtail occurs in Arizona and Mexico.

PALE SWALLOWTAIL occurs all along the West Coast to the eastern slope of the Rockies and is quite common locally. Larva feeds chiefly on buckthorn. At least two broods occur annually.

ZEBRA SWALLOWTAIL, an eastern species more common in the South, varies in marking and size. Spring forms are smallest; later broods larger with longer tails. Larva feeds on pawpaw.

3.8-4.5"

3.5-4.0"

28

larva

PARNASSIUS is a more primitive genus than *Papilio*, the true swallowtails. The larva shows many habits of skippers (pp. 74-80) and, like them, is covered with short hairs. It lacks the scent horns of swallowtails. The pupa is not like those of swallowtails either, but, like those of skippers, is smooth and brownish, and is formed in a leafy shelter on the ground or in grass. The adult parnassius is not like swallowtails in shape or coloration but is pale white or yellowish, with markings that vary greatly. Parnassians occur mostly in the mountains, where adults are on the wing by midsummer. Female bodies lack the hairiness of the male.

CLODIUS, a variable species, is distinguished from Smintheus by its black antennae. It lacks the red spots often found on the forewings of Smintheus.

SMINTHEUS differs from other parnassians in having white antennae with black rings. The larva feeds on stonecrop *(Sedum)* and saxifrage *(Saxifraga)*.

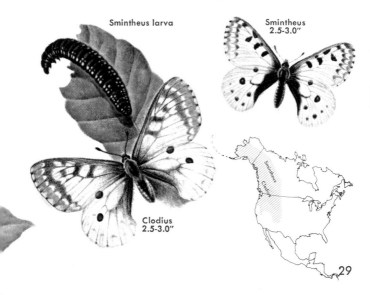

Smintheus larva

Smintheus
2.5-3.0"

Clodius
2.5-3.0"

SULPHURS AND WHITES form a world-wide family of several hundred species, including many species in temperate parts of the Northern Hemisphere. They are among the first butterflies to appear in spring. Nearly all are yellow, orange, or white. Females differ from males in pattern and often in color. The butter-yellow color of European sulphurs probably suggested the name *butterfly*. Some are often seen around the edges of puddles. Eggs are spindle-shaped, sculptured with fine ridges and pits. The larva, usually long, green, and slender with little hair, feeds mainly on legumes and mustards. Some are crop pests. The pupa, often compressed and triangular, is held in place by a silk girdle. Most species have more than one brood a year, especially in the South, where three or even more may occur.

ALFALFA BUTTERFLY, also called the Orange Sulphur, occurs in many hybrid forms, crossing with the Clouded Sulphur. It can be distinguished from the Clouded Sulphur by the orange of the uppersides. The undersides of both are similar. Some females are white. The larva is a pest of alfalfa. Several broods each year. The butterfly ranges from Canada to Mexico.

male
1.8-2.0"

female

female

male
1.3-2.3"

white female

larva

pupa

CLOUDED SULPHUR, also called the Common Sulphur, ranges through most of North America but is most common in the East. It is the "puddle butterfly" that swarms in moist places and over clover fields. Like the Alfalfa Butterfly, some females are white, but generally the Clouded female has less black on the margin of the wings. The larva is more common on clover than on alfalfa and cannot be distinguished from that of the Alfalfa Butterfly. There are several broods yearly.

PINK-EDGED SULPHUR can be distinguished from other sulphurs by the pink wing edges and the pink-edged silver spot on the underside of each hindwing.

1.3-2.0"

Alfalfa

Clouded

31

Sara Orange Tip
1.0–1.3"

male

female

SARA ORANGE TIP is variable. The underside of the hindwings has an irregular "mossy" appearance from the greenish marbling. The amount of marbling is reduced in the second of the two annual generations. Larva feeds on wild mustards.

OLYMPIA MARBLE, named for the pronounced green marbling on the underside of the hindwing, is closely related to the orange tips. The larva feeds mostly on Hedge Mustard. The Olympia Marble produces only one brood a year, in the spring.

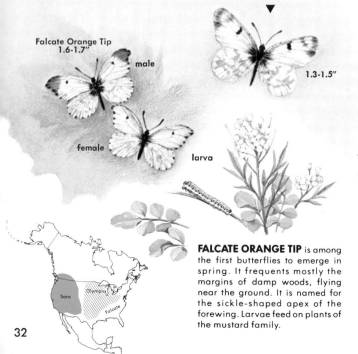

Falcate Orange Tip
1.6–1.7"

male

female

1.3–1.5"

larva

FALCATE ORANGE TIP is among the first butterflies to emerge in spring. It frequents mostly the margins of damp woods, flying near the ground. It is named for the sickle-shaped apex of the forewing. Larvae feed on plants of the mustard family.

32

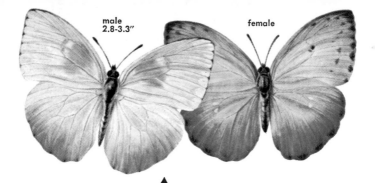

male
2.8-3.3"

female

▲

ORANGE-BARRED SULPHUR is common along the Gulf of Mexico, occasionally straying into middle Atlantic and midwestern states. The larva, yellowish-green, with black and yellow bands and small black spines, feeds on Cassia and on other closely related plants of the pea family. At least two broods each year.

CLOUDLESS SULPHUR, also known as the Giant Sulphur, is abundant in the tropics and common in our southern states. Huge flocks during migration are an impressive sight. Breeding is continuous in the tropics, but to the north there are two broods with adults overwintering. Wild Senna is its chief food.

▼

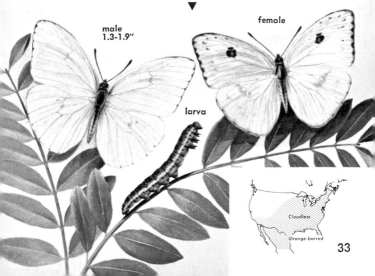

male
1.3-1.9"

female

larva

Cloudless

Orange-barred

33

male
1.8-2.0"

female

▲
CALIFORNIA DOG-FACE can be distinguished from the Southern Dog-face by the lack of dark margins on the upperside of the hindwing. It sometimes strays eastward from its normal range.

The larva feeds on False Indigo. Two broods develop yearly; the adults are on the wing in spring and midsummer. The name dog-face comes from the "poodle face" marking on the forewing.

SOUTHERN DOG-FACE, noted for its rapid flight, is more common in the southern part of its range. Like the California Dog-face, it is double-brooded. The larvae, highly variable in markings, feed on Wild Indigo and also on various clovers.

2.3-2.5"

◄

1.3-1.6"

larva

Little Sulphur

pupa

▲
LITTLE SULPHUR is common east of the Rockies. The larva feeds on Senna, Partridge Pea, and other legumes. There are two or three broods a year. This species migrates in large flocks.

34

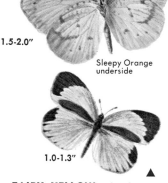

1.5-2.0"

Sleepy Orange underside

SLEEPY ORANGE is so named because it is slower in flight than other sulphurs. It is very common in the South, where it breeds throughout most of the year. Males frequently congregate in large numbers at puddles. The velvety green larva with a yellowish stripe on each side feeds mainly on Senna.

1.0-1.3"

FAIRY YELLOW, also known as Barred Sulphur, has a gray bar in the forewing of males and some females. It ranges from Florida and Texas southward, feeding on Joint Vetch and other legumes.

0.8-1.3"

DAINTY SULPHUR occurs from Ga. to so. Calif. and up the Miss. Valley to the Great Lakes. Females are heavily marked with black. It is double-brooded.

1.3-1.6"

MUSTARD WHITE has a circumpolar range. The veins on the underside are outlined with dark scales. The larva, green with greenish-yellow stripes, feeds on various mustards.

California

Southern

Mustard

Sleepy

35

1.5-2.0"   1.8-2.3"

PINE WHITE, a pest of pines and Balsam Fir in the West, has one brood and overwinters as eggs. The female has more black markings than the male.

GIANT WHITE is common in the tropics and breeds to southern Texas, straying northward. Like the Great Southern White it also has a dark phase.

GREAT SOUTHERN WHITE of the Gulf Coast and Miss. Valley, sometimes migrates. If so, a dark phase is involved. The larva feeds mostly on mustards.

FLORIDA WHITE is a widespread butterfly that strays north from Fla. and Tex. All have orange on the undersides; most females also have dark marks.

2.5-2.8"   1.6-2.3"

female and larva

male
1.3-1.6"

CHECKERED WHITE, or Common White, occurs all over temperate N.A. It was more common before the Cabbage Butterfly arrived and spread. Larva feeds on cabbage and other mustards. Adults occur early in spring and produce at least three broods.

CABBAGE BUTTERFLY, introduced from Europe about 1860, has spread across N.A. and become a pest of cabbage, broccoli, kale, cauliflower, and other mustards and of the garden nasturtium. It is one of the first butterflies to emerge in spring.

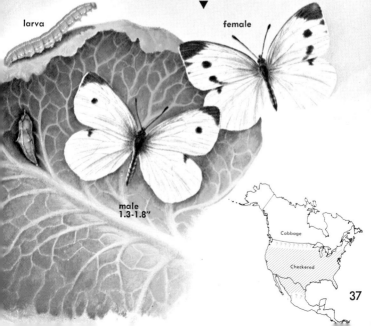

larva

female

male
1.3-1.8"

Cabbage

Checkered

BRUSH-FOOTED BUTTERFLIES are named for their tiny forelegs, useless for walking, hairy in males.

MILKWEED BUTTERFLIES, chiefly tropical, number only two species in North America, both common.

3.5-3.9"

THE MONARCH, one of the best known butterflies, is noted for its migratory habits. In fall, flocks of Monarchs move southward to California and Mexico. Resting migrants or winter residents may cover entire trees. In spring they return northward to their breeding areas, some as far as southern Canada. Three or four broods may be produced in one year. The male scent glands are marked by a spot of dark scales in the center of the hindwings; this spot is not found on the female. Females differ also in having broader black vein lines. The larva feeds on milkweeds and related plants, the juices of which cause the Monarch's unpalatability to many birds. The Monarch's development takes about a month from conical eggs—laid singly on leaves or blossoms—to adult, which emerges from a shiny green, gold-speckled hanging chrysalis. The larva, striped with yellow, black, and white, is about 2 in. long when fully grown.

3.1-3.3"

THE QUEEN resembles the Monarch but is smaller and darker brown. The brownish larva with brown and yellow transverse stripes feeds on milkweeds. There are three broods yearly. A subspecies called Bates' Queen has the veins of the upperside of the hindwing edged with grayish white. Bates' Queen is found in southern Arizona, New Mexico, and Texas.

Monarch pupa

larva

Monarch

Queen

39

SATYRS are butterflies of rather dull color, usually brown or gray with eyespots on both upper and under sides. Most prefer woods or woods margins. Satyr larvae have forked tails, feed at night on grasses, and overwinter as tiny larvae.

EYED BROWN has a weak dancing flight. It frequents damp meadows with tall grasses and also the margins of woods. Local colonies are found east of the Rockies from Canada to Florida.

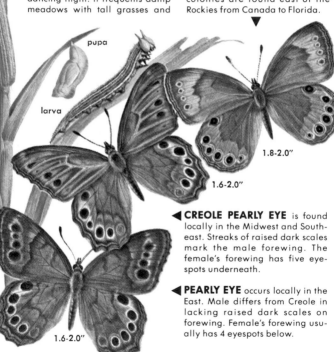

pupa

larva

1.8-2.0"

1.6-2.0"

◄CREOLE PEARLY EYE is found locally in the Midwest and Southeast. Streaks of raised dark scales mark the male forewing. The female's forewing has five eyespots underneath.

◄PEARLY EYE occurs locally in the East. Male differs from Creole in lacking raised dark scales on forewing. Female's forewing usually has 4 eyespots below.

1.6-2.0"

NORTHERN PEARLY EYE ranges from Quebec so. thru Appalachians to n. Ga., and so. from Manitoba to Ark. and Miss. Differs from Pearly Eye by having orange, not black, antennal knobs.

Little Wood
1.8-2.0"

underside

Gemmed
1.8"

Carolina
1.5"

Georgia
1.7"

LITTLE WOOD SATYR prefers open woods and meadows overgrown with shrubbery. Occurs east of the Rockies.

CAROLINA SATYR is mousegray above, without eyespots. Occurs from N.J. to Fla., west to Texas, and up the Miss. Valley.

GEMMED SATYR is marked by a prominent violet-gray patch under its hindwing. Virginia to Illinois, south to Fla. and Mex.

GEORGIA SATYR prefers marshy areas or open pine woods. Distribution is similar to that of the Carolina Satyr.

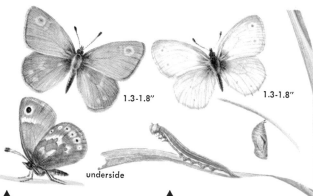

1.3-1.8"

1.3-1.8"

underside

▲
PLAIN RINGLET ranges through Canada south to Conn. and the northern Midwest. The underside of the hindwing has an isolated light-colored patch.

▲
CALIFORNIA RINGLET is very common west of the Rockies. It is near-white above, darker below, with an angular white patch on its hindwing.

1.5-1.8"

1.8"

RIDING'S SATYR at rest is the color of mossy rock and thus is well camouflaged in its environment. Pupa forms underground.

COMMON ALPINE is common in spring in the mountains from New Mexico to Wash. and Alaska. Larva feeds on grasses.

COMMON WOOD NYMPH varies so in color and pattern in its nation-wide range that different forms are hard to recognize. The northeastern form is illustrated. Southeastern form is larger, and the lower eyespot on the forewing is smaller. In the West it is darker and smaller, and does not have the light band on the forewing.

2.0-2.8"

Common Wood Nymph

pupa

larva

NEVADA ARCTIC is also called the Greater Arctic. The female lacks the dark shading on the male's forewing. Other species of arctics, mostly smaller and some without eyespots, occur in Canada, the Rockies, and alpine New England.

2.3-2.5"

42

HELICONIANS are peculiar to the American tropics. They are reputed to be protected against predation by their unsavory taste and odor. Some are mimicked by other butterflies. The forewings are twice as long as wide. The eggs are rounded, and about twice as long as wide. The larvae, with rows of long branched spines, feed on leaves of the Passion Flower. The pupae are unusually angular.

2.6-2.8"

GULF FRITILLARY ranges through South America to New Jersey and Iowa. In color and wing shape it resembles the greater fritillaries (p. 44-46).

JULIA, native to South and Central America, is a strong flier. Occasionally it appears in large swarms in southern Texas or Florida. The female is much lighter and duller than the male. The chrysalis lacks the spines of the Zebra.

3.2-3.8"

ZEBRA, found in or near woods, is a weak flyer and moves about slowly. Zebras gather in colonies at night. Males are attracted to chrysalids of females just before the latter emerge. Strays from tropical America to South Carolina and Kansas.

larva

pupa

3.0-3.4"

GREATER FRITILLARIES are common in temperate regions. Caterpillars feed at night, mostly on violets. Most species are single-brooded, overwintering as tiny larvae.

VARIEGATED FRITILLARY ranges over most of the U.S. except the Pacific Northwest. It lacks the typical silver spots on the underside of the wings. The Mexican Fritillary, similar in form but with plainer hindwings, is found from southern Texas to southern California.

REGAL FRITILLARY frequents roadsides and wet meadows, feeding on milkweeds and thistles. Both rows of spots on the hindwing of the female are white but only the inner row of the male is white. Larva is like Great Spangled, but black, mottled with yellow. Formerly fairly common throughout the Northeast as far west as Nebraska and Missouri, this striking species now appears to be diminishing alarmingly in numbers.

Variegated
Fritillary
2.3-2.5"

Regal Fritillary
3.4-3.6"

pupa

larva

pupa

3.3-3.8"

underside

GREAT SPANGLED FRITIL-LARY is single-brooded in the North and double-brooded in the South. Larva hibernates soon after hatching and the following spring feeds at night on violets. The adults, in typical fritillary fashion, prefer marshes and damp meadows. This is one of the best known fritillaries.

APHRODITE, similar to Great Spangled, is smaller and has a narrower, yellowish marginal band under the hindwing. It prefers high elevations from so. Can. to Ga. w. to the Rockies. Another species, Atlantis, is like Aphrodite, but its forewing has a dark margin. It occurs from Can. so. thru mtns. to Virginia.

Great Spangled

Regal

3.0-3.3"

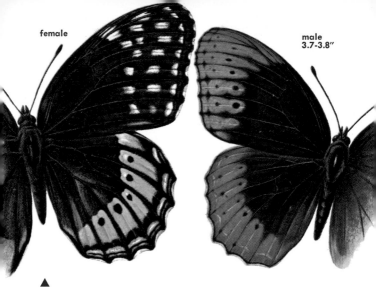

female

male
3.7-3.8″

▲

DIANA is an unusual fritillary because the sexes differ so in color and markings and because it prefers woodlands to open country and is more attracted to manure piles than to flowers. It ranges from the southern Appalachian Mts. west to Illinois.

NEVADA FRITILLARY is quite common in sections of the Rockies and foothills of the Sierra Nevadas. Greenish underside of hindwing is a good identifying characteristic.

EURYNOME resembles the Nevada Fritillary, but the greenish tint covers only top third of underside of the hindwing. Occurs in the Rockies from N. Mex. to Can., west to Cascades.

▼

2.5-3.5″

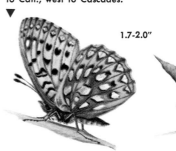

1.7-2.0″

LESSER FRITILLARIES resemble greater fritillaries but are smaller, and most lack the silver spots under the hindwing. The larvae of most species feed on violets.

1.5-2.0"

underside

1.6-1.8"

underside

▲
SILVER-BORDERED FRITILLARY appears in many varieties. Some are also found in Europe. All have the heavily silvered underside of the hindwing.

▲
EASTERN MEADOW FRITILLARY has outer margin angled near apex, not curved as in other fritillaries. It lacks the dark outer margin, but has black spots.

◀ **WESTERN MEADOW FRITILLARY** resembles the Silverbordered in shape but is different beneath. Those in the North are darker above than those in the South. They are common in mountain valleys of the West, ranging from Colorado to California and north to British Columbia.

1.5-1.9"

underside

THE CHECKERSPOTS, small- to medium-sized, lay their eggs in groups. The spiny caterpillars feed together for a while. The free-hanging pupa is whitish with dark blotches.

1.8-2.6"

(1)

(2)
1.5-1.6"

(3)
1.4-1.7"

(3) underside

(4)

(1)
larva

(1)
pupa

(1) THE BALTIMORE, though widespread, is local, seldom found far from its wet meadow food plant, Turtlehead. Many variations occur. Several species are found in the West.

(2) SILVERY CHECKER-SPOT is similar to Harris' on the upper surface. Found along roads, lakes, and open meadows from Maine to North Carolina and west to the Rockies.

(3) HARRIS' CHECKER-SPOT is also variable and very local. It prefers damp fields and underbrush. Ranges from Nova Scotia west to Manitoba and south to Illinois and W. Va.

(4) CHALCEDON CHECK-ERSPOT is quite variable in color and pattern. It is common along the Pacific in the lower mountain levels and feeds on plants of the figwort family.

Chalcedon

Baltimore

1.8-3.0"

48

THE CRESCENTS are closely allied to checkerspots but are smaller. Named for pale crescent on outer margin of the hindwing, beneath. The larvae stay together but do not make webs. Summer brood is paler than the winter brood.

(1) MYLITTA CRESCENT, abundant on the West Coast, varies greatly. It generally shows a pale crescent both above and beneath. Larva feeds on thistles.

(2) FIELD CRESCENT is common in California and occurs from Arizona to Alaska, with both spring and fall broods, in damp meadows and along streams. Larva feeds on asters.

(3) PHAON CRESCENT is easy to tell from other crescents by its underside. The forewing above has a band of white to yellowish spots. Gulf states to Calif.

(4) PEARL CRESCENT, one of the most common butterflies, is found around puddles and flowers. It ranges over all North America south of Hudson Bay, except for the Pacific Coast.

underside

(1) 1.2-1.5″

(2) 1.2-1.4″

(3) 0.9-1.3″

(4) 1.3-1.7″

light female

(4)

dark female

larva (4)

pupa

ANGLE WINGS are named for the sharp, angular margins of their wings. The undersides of the wings closely resemble dead leaves or bark, camouflaging angle wings in their woodland haunts. Eggs sometimes occur in a hanging chain. The larva is spiny; the angular pupa hangs free. Like crescents, angle wings have light and dark seasonal forms. They hibernate as adults.

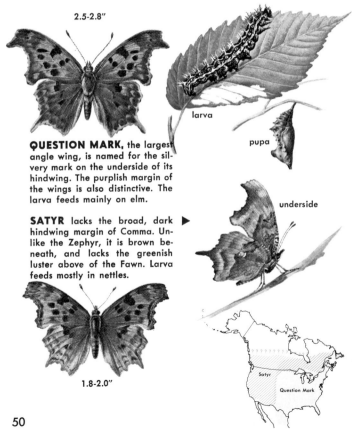

2.5-2.8"

larva

pupa

underside

QUESTION MARK, the largest angle wing, is named for the silvery mark on the underside of its hindwing. The purplish margin of the wings is also distinctive. The larva feeds mainly on elm.

SATYR lacks the broad, dark hindwing margin of Comma. Unlike the Zephyr, it is brown beneath, and lacks the greenish luster above of the Fawn. Larva feeds mostly in nettles.

1.8-2.0"

Satyr

Question Mark

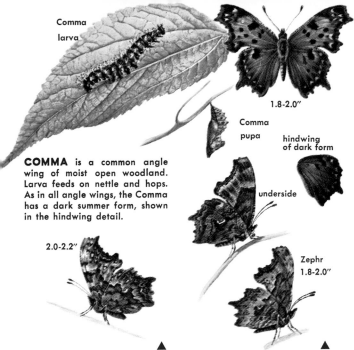

Comma
larva

1.8-2.0"

Comma
pupa

hindwing
of dark form

underside

Zephr
1.8-2.0"

COMMA is a common angle wing of moist open woodland. Larva feeds on nettle and hops. As in all angle wings, the Comma has a dark summer form, shown in the hindwing detail.

2.0-2.2"

FAWN, or Green Comma, has a greenish tint to its wings. Found in the mountains from eastern Canada and Carolinas to the N.W. states. Larva lives on birch and alder, feeding on the undersides of leaves.

ZEPHYR, like the Satyr, lacks dark margin on hindwing but is gray beneath. Seen from May to Sept. Larva feeds on elm and currant. The darker form below was once considered a separate species—*silenus.*

▼
dark form

51

BUCKEYES are brightly colored butterflies, all of which have a large eyespot on the upperside of both the hind and the fore wings. There are some fifty species throughout the world, but only one of them is common in North America.

(1)

2.0-2.3"

underside

(7)

(2)

(3)

(4)

(5)

(6)

Buckeye

Painted Lady

THE BUCKEYE, a variable species, overwinters as an adult. Larva feeds mainly on plantain and Gerardia. In its development, larva (1) attaches itself to a support (2) and becomes a pupa (3). Adult develops in the pupa (4) and emerges with soft wings (5) which soon expand (6) and dry (7).

THISTLE BUTTERFLIES are a widespread group. One species, the Painted Lady, ranges through all temperate and some tropical areas. These butterflies frequent flowers, especially thistles. Adults hibernate. Some species migrate. The larvae are spiny.

Painted Lady
2.0-2.3"

underside

West Coast Lady
2.0-2.3"

underside

PAINTED LADY is called the Cosmopolitan because of its wide range. It is also noted for its migrations. The larva builds a webbed nest on the food plant, usually thistle. Adults prefer open places. There are usually two broods a year in the North.

WEST COAST LADY ranges from the Rockies westward and south to Argentina. It differs from the Painted Lady in lacking the white bar on the upper surface of the forewing. It is easily captured while feeding on flowers. The larva feeds on mallows.

2.0-2.3"

underside

larva

pupa

▲

RED ADMIRAL is found world-wide in north temperate regions. It is a swift erratic flier seen in open woodland and around Butterfly Bush. The larva lives and feeds singly on leaves of nettles, the edges of which it draws to-gether with silk. Adults hiber-nate. The Red Admiral is double-brooded; the second brood is larger and darker than one shown here. The ranges of the Red Admiral and American Painted Lady are almost alike.

2.0-2.3"

Red Admiral and Amer. Painted Lady

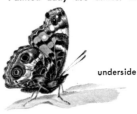

underside

AMERICAN PAINTED LADY, or Hunter's Butterfly, has two large eyespots on the underside of the hindwing. Painted and West Coast Ladies have 5 small spots each. Greenish eggs are laid on everlasting and burdock. Larva is black with yellow stripes.

TORTOISE SHELLS include butterflies that resemble angle wings, but the inner margin of the forewing is straight instead of concave. Adults hibernate and may be seen very early in spring. Eggs are laid in clusters. Tortoise shells are a circumpolar group which is widespread in the Northern Hemisphere.

COMPTON TORTOISE SHELL frequents open woodlands, where its undersides provide excellent camouflage.

MILBERT'S TORTOISE SHELL is a northern species of open areas and mountain meadows. Larva feeds on nettle.

MOURNING CLOAK occurs throughout temperate N.A. Adults that have hibernated may be seen sunning in early spring with open wings. Eggs are laid in masses around the twigs of elm, willow and poplar. Larvae may become pests.

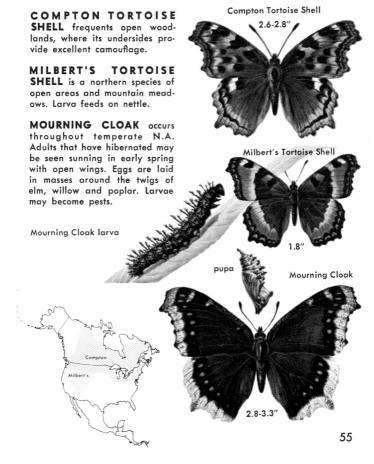

Compton Tortoise Shell
2.6-2.8"

Milbert's Tortoise Shell
1.8"

Mourning Cloak larva

pupa

Mourning Cloak
2.8-3.3"

Compton

Milbert's

ADMIRALS AND SISTERS total six species in N.A. The larva, not as spiny as those of other brush-footed butterflies, feeds on a variety of trees. These species are mostly double-brooded, and the tiny larva hibernates in silken shelters on the food plant.

WHITE ADMIRAL, or Banded Purple, lives in upland hardwood forests and on mountain slopes, where its larva feeds mostly on birch, willow, and poplar. The adult is readily attracted by putrid barnyard odors.

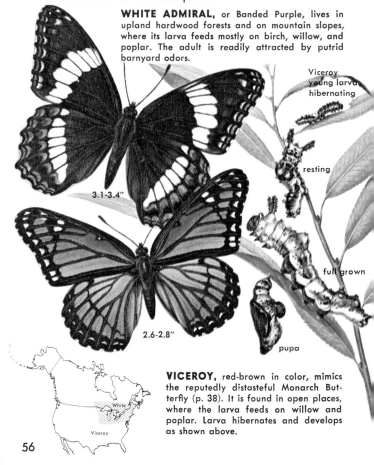

Viceroy young larva hibernating

resting

3.1-3.4"

full grown

2.6-2.8"

pupa

VICEROY, red-brown in color, mimics the reputedly distasteful Monarch Butterfly (p. 38). It is found in open places, where the larva feeds on willow and poplar. Larva hibernates and develops as shown above.

White

Viceroy

56

WEIDEMEYER'S ADMIRAL has white spots along the margin of the forewing. Found on mountain slopes and wet places where aspen and willow grow.

LORQUIN'S ADMIRAL has an orange tip to forewing and a white band on both wings. Found in river bottomland. Larva feeds on cherry, willow, and poplar.

3.0-3.3"

2.3-2.8"

3.1-3.4"

RED-SPOTTED PURPLE, considered by some to be a subspecies of the White Admiral, prefers lower altitudes and a warmer climate than that species. It also prefers more open areas, where the larva feeds mostly on Wild Cherry.

2.5-3.0"

CALIFORNIA SISTER is similar to Lorquin's Admiral but has blue lines on undersides of wings. It is a common California butterfly, frequenting the upper branches of live oaks, on which the larva feeds. The butterfly rarely sips nectar from flowers.

LEAFWING BUTTERFLIES are a tropical group in which the undersides of the wings resemble dead leaves. Color and wing shape vary greatly. Two seasonal forms occur—a wet and a dry. The forewings of the dry-season form are less curved. Larva hides by day in a rolled leaf. The goatweeds are the only North American species.

1.9-2.2"   1.9-2.2"

GOATWEED BUTTERFLY ranges from Ga. and Tex. up the Miss. Valley. Its dry-season form is lighter in color. Female is like Morrison's, but light spots on wings form a continuous band.

MORRISON'S GOATWEED has a tropical range but enters Texas. Male is quite similar to Goatweed Butterfly but is more brilliant. Female (illustrated) differs in color and pattern.

EMPEROR, OR HACKBERRY, BUTTERFLIES are found near Hackberry trees, on which the larva feeds in colonies. The striped caterpillar tapers toward both ends and bears two "horns" behind the head. It hibernates when partly grown. Adults show much geographic variation. The females are larger than the males and are also lighter in color.

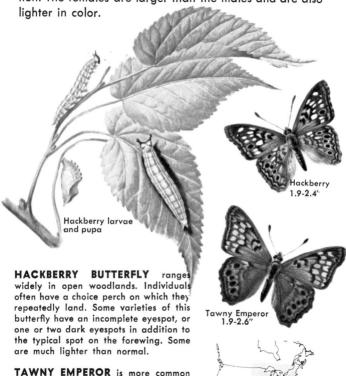

Hackberry larvae and pupa

Hackberry
1.9-2.4"

Tawny Emperor
1.9-2.6"

HACKBERRY BUTTERFLY ranges widely in open woodlands. Individuals often have a choice perch on which they repeatedly land. Some varieties of this butterfly have an incomplete eyespot, or one or two dark eyespots in addition to the typical spot on the forewing. Some are much lighter than normal.

TAWNY EMPEROR is more common in the South, though it ranges north to New England. Lacks dark eyespots on the forewing. Larva similar to Hackberry Butterfly's but has branched head spines.

59

Florida Purple Wing 1.6-2.0"

2.6-2.8"

Ruddy Dagger Wing

PURPLE WINGS are tropical butterflies, usually dull purplish above and well marked on the undersides. Two species occur in southern Florida and Texas.

DAGGER WINGS are mainly tropical butterflies, with prolonged tips to their forewings, resembling small swallowtails. One species breeds in the United States.

FLORIDA PURPLE WING occurs in dense hardwood hammocks. Dingy Purple Wing (not shown) is slightly smaller and lacks most of the purple sheen

RUDDY DAGGER WING of southern Fla. and Texas, may stray northward. The ornate filament-bearing larva feeds on fig and *Anacardium*.

TROPIC QUEENS are tropical butterflies noted for their beauty and the females' trait of mimicking milkweed butterflies (p. 38). The Mimic is a species that was probably introduced into the American tropics from the Old World a long time ago. The Mimic occurs in the West Indies and locally in Florida.

Mimic male 2.5-3.0"

Mimic female

60

METALMARKS are small butterflies usually having metallic spots, from which the common name is derived. Many of the fifteen plainly colored species occurring north of Mexico are difficult to distinguish. In the tropics, metalmarks are common and occur in many different bright color patterns. Males have four walking legs, females six. They rest with wings outstretched. The larvae resemble those of hairstreaks. The pupa is hairy, suspended by a stem, and supported by a silk thread.

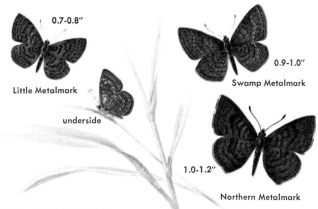

0.7-0.8"

Little Metalmark

underside

0.9-1.0"

Swamp Metalmark

1.0-1.2"

Northern Metalmark

LITTLE METALMARK is more common in the southern part of its range. It occurs in open grassy areas, where it is distinguished by its small size and its uncheckered wing margins.

NORTHERN METALMARK is relatively rare and has been confused with similar species. The wings are darker than those of the Little and the Swamp and have an irregular dark band. The Northern prefers dry hilly terrain and open woods.

SWAMP METALMARK lacks the inner dark irregular band of the Northern, and wing margins are slightly checkered. It occurs in wet meadows and swamps in summer. Overwinters as larva that feeds on swamp thistle.

Swamp
Northern
Little

61

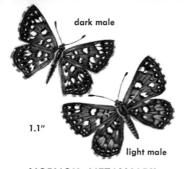

dark male

underside

1.1"

light male

1.0-1.3"

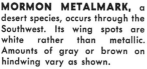

MORMON METALMARK, a desert species, occurs through the Southwest. Its wing spots are white rather than metallic. Amounts of gray or brown on hindwing vary as shown.

NAIS METALMARK occurs from Colorado to Mexico. Its wing fringes are checkered, but in overall appearance it is not distinctly like other metalmarks. The larva feeds on Wild Plum.

SNOUT BUTTERFLIES are easily recognized by the long projecting mouth parts (palpi) which resemble snouts. Like the metalmarks, males have four walking legs and the females six. The Common Snout Butterfly is the only snout butterfly regularly occurring north of Mexico. The larva, which grows very rapidly, feeds on Hackberry.

underside

larva

Snout

Common Snout Butterfly
1.8-2.0"

GOSSAMER WINGS are small- to medium-sized butterflies, often with hairlike tails on the hindwings. They are usually blue, coppery, gray, or dull brown above.

HAIRSTREAKS, about 70 species north of Mexico, have a swift, darting flight and are readily attracted to flowers.

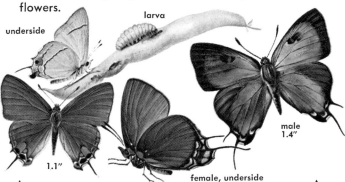

underside

larva

male
1.4"

1.1"

female, underside

▲
GRAY or COMMON HAIRSTREAK is also called the Cotton Square Borer or the Bean Lycaenid because of damage it sometimes does to crops. It overwinters in the pupal stage and emerges early in the spring.

▲
GREAT PURPLE HAIRSTREAK females have two tails on each hindwing, as do some males. The female lacks sex-pads—black spots on forewing. The larva feeds on mistletoe. Double-brooded.

◀ **COLORADO HAIRSTREAK** is actually more purple in color than the Great Purple. The underside has a typical banded pattern. This species is commonly found around scrub oaks.

1.5"

63

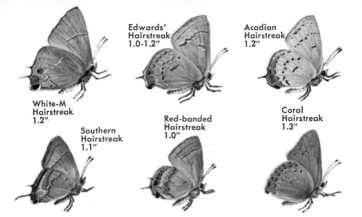

White-M Hairstreak
1.2"

Edwards' Hairstreak
1.0-1.2"

Acadian Hairstreak
1.2"

Southern Hairstreak
1.1"

Red-banded Hairstreak
1.0"

Coral Hairstreak
1.3"

WHITE-M HAIRSTREAK, a southeastern species, is named for the inverted white M on the underside of the hindwing. The upper surface of the wing is blue.

ACADIAN HAIRSTREAK has widely separated spots instead of transverse lines. The larva feeds on willow. Adults are found in wet areas where willows grow.

RED-BANDED HAIRSTREAK occurs from Florida and Mexico to New York and Michigan, but is commoner in the South. Male upperside is brown; female, bluish.

EDWARDS' HAIRSTREAK has oval spots that form broken transverse lines. It frequents thickets of Scrub Oak, on which the larva feeds.

SOUTHERN HAIRSTREAK has orange patches on the uppersides of both wings, larger on the hindwing. It is single-brooded. The larva feeds on oak.

CORAL HAIRSTREAK is tailless. Coral red spots on underside may form a solid band. It overwinters in the egg stage. Adults appear by midsummer.

California Hairstreak 1.1-1.3"

Banded Hairstreak 1.0-1.2"

Hedgerow Hairstreak 1.2"

Striped Hairstreak 1.2"

Olive Hairstreak 0.9-1.0"

Sylvan Hairstreak 1.0"

CALIFORNIA HAIRSTREAK is single-brooded, appearing on the wing in midsummer in the foothills. Larva feeds on *Ceanothus* and, probably, on oak.

HEDGEROW HAIRSTREAK, reddish brown above, is common in the Rockies and west to the Pacific coast in summer. Feeds on *Cercocarpus* and *Ceanothus.*

OLIVE HAIRSTREAK, double-brooded, overwinters as a pupa. The adults occur in spring and midsummer, usually near red cedars, the larval food plants.

BANDED HAIRSTREAK occurs in late spring and early summer, usually in or near woodlands. It overwinters in the egg stage. The larva feeds on oak and hickory.

STRIPED HAIRSTREAK is distinctly striped underneath. It is widely distributed east of the Rockies. The larva feeds on many plants, including oak and willow.

SYLVAN HAIRSTREAK resembles California Hairstreak but is lighter beneath and has only one small red spot. The larva feeds on willow.

Banded Elfin male 0.8-1.2"

Western Banded Elfin 1.2" underside

underside

larva

Brown Elfin 0.9-1.0"

Banded Elfin female

ELFINS are small- to medium-sized brown butterflies; females are larger and less drab than males. Elfins overwinter as pupae and have only a single brood yearly. They are among the first butterflies to appear in spring. The males of all elfins but Henry's have a "sex-spot" on the upper side of the forewings.

WESTERN BANDED ELFIN resembles the Banded Elfin, but the band (middle of forewing, underside) is less irregular. Larva feeds on pine.

BANDED ELFIN, also called Pine Elfin, is usually found in open pine stands. The larva feeds primarily on the seedlings of both hard and soft pines, on which it is well camouflaged.

BROWN ELFIN, reddish brown on the underside, is found in and along the edges of open woods where its food plants, blueberry and Sheep Laurel, grow.

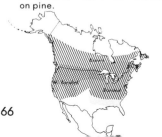

66

Western Elfin
0.9-1.0"

Hoary Elfin
0.9-1.0"

Henry's Elfin
0.9-1.1"

Frosted Elfin
0.9-1.5"

WESTERN ELFIN is obscurely marked beneath. It occurs in both lowlands and mountains, often frequenting *Ceanothus* blossoms. The larva feeds on sedum.

HOARY ELFIN gets its name from the gray color on the underside. It occurs in open, dry, heath-covered areas. The larva feeds on bearberry.

HENRY'S ELFIN of open woods is less gray on the underside and is dark brown at the base of the scalloped hindwing. The larva feeds on blueberry.

FROSTED ELFIN male has a sex-spot on the upperside of forewing. Hindwing has more scalloped border and less color contrast than that of Henry's Elfin.

67

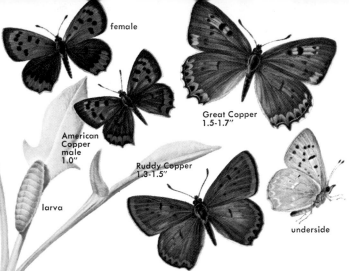

female

Great Copper
1.5-1.7"

American
Copper
male
1.0"

larva

Ruddy Copper
1.3-1.5"

underside

COPPERS occur chiefly in the Northern Hemisphere, with about sixteen species in the United States and Canada. Most are reddish or brown and have a coppery luster, but one, the Blue Copper, is bright blue. Most species frequent open areas and roadsides.

AMERICAN COPPER occurs from spring to fall in fields where Sheep Sorrel, food plant of the larva, grows. Overwinters in the pupal stage. Adults in spring are brighter and less spotted.

GREAT COPPER is one of the largest of the coppers. Males have fewer black dots and less orange on the wing margins. The females feed on dock. Adults emerge in summer.

RUDDY COPPERS have white margins on the wings and fewer black spots on the underside of the hindwing than other coppers. The female resembles the American Copper but is larger and not as brightly colored. Feeds on Arnica.

male

Purplish Copper
1.2-1.3"

female

underside

Gorgon Copper
1.3-1.4"

male   Bronze Copper   female
1.3-1.4"

PURPLISH COPPER is common from spring to fall, mostly in moist meadows. The underside of the hindwing is marked with a faint red line. The larva feeds mostly on dock and knotweeds.

BRONZE COPPER frequents wet meadows. It is double-brooded and hibernates in the egg stage. The margin of the underside of the hindwing has a broad orange band. The larva feeds mostly on several species of dock and knotweed.

GORGON COPPER has only a midsummer brood. Its underside is typical of the coppers. The female resembles that of the Purplish Copper but is less bright. The larva feeds on *Eriogonum*.

69

1.3"

larva

pupa

The Harvester

THE HARVESTER occurs only in North America, but a few close relatives are found in Africa and Oriental tropics. The larva feeds on woolly aphids that live on alder, beech, and witch hazel, and becomes full grown in as little as ten days. There are several generations a year. Winter is passed in the pupal stage. The markings of the pupa resemble a monkey's face.

BLUES are small and usually blue above. The larvae of some species secrete "honeydew" and are attended by ants for this liquid. There is much seasonal variation in color. The sexes differ; females usually are darker, with wider dark borders on the upperside of the wings.

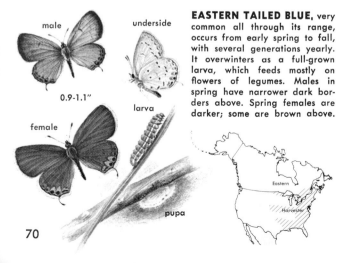

male

underside

0.9-1.1"

larva

female

pupa

Eastern

Harvester

EASTERN TAILED BLUE, very common all through its range, occurs from early spring to fall, with several generations yearly. It overwinters as a full-grown larva, which feeds mostly on flowers of legumes. Males in spring have narrower dark borders above. Spring females are darker; some are brown above.

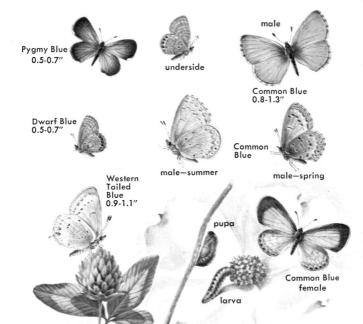

Pygmy Blue
0.5-0.7"

underside

male

Common Blue
0.8-1.3"

Dwarf Blue
0.5-0.7"

Common
Blue

male—summer

male—spring

Western
Tailed
Blue
0.9-1.1"

pupa

larva

Common Blue
female

PYGMY BLUE and Dwarf Blue are the smallest of all North American butterflies. The Pygmy is common in its range. Its larva is well camouflaged on Lamb's Tongue, its food plant.

DWARF BLUE is similar but lacks the white spot and fringe on the upperside of the forewing.

COMMON BLUE, or Spring Azure, occurring throughout North America, is another early spring butterfly. Spring forms are darker than summer forms, with spots on the undersides sometimes fused. The underside markings of summer forms are usually pale. The slug-like larva feeds on flowers and excretes a sweet liquid called honeydew, for which it is followed by ants.

WESTERN TAILED BLUE is most easily distinguished from Eastern by its less-spotted underside. In some areas the Eastern and Western Tailed Blues occur together.

71

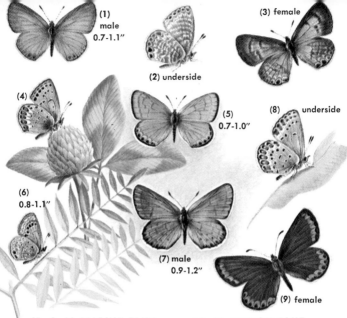

(1) male 0.7-1.1"

(2) underside

(3) female

(4)

(5) 0.7-1.0"

(8) underside

(6) 0.8-1.1"

(7) male 0.9-1.2"

(9) female

(1, 2, 3) MARINE BLUE appears later in spring than other blues. Larva feeds mostly on buds and blossoms of wisteria, alfalfa, locoweed, and other legumes.

(4, 5) ACMON BLUE occurs early spring to fall. The female is brownish or bluish. The broad orange band with black spots on hindwing is distinctive.

(6) REAKIRT'S BLUE is easily distinguished by the white-ringed black spots on the underside of the forewing. Mesquite is one of its food plants.

(7, 8, 9) ORANGE-BOR-DERED BLUE, or Melissa Blue, is double-brooded. Larva feeds on legumes. Note orange spots on upper hindwing of female.

(1) 0.9-1.2"

(2) female

(3) underside

(4) male 0.9-1.2"

(5) female

(6) underside

(7) male 0.9-1.0"

(8) female

(9) underside

(10) male 1.0-1.1"

(1) SILVERY BLUE lacks the black spots along the margin of underside of hindwing. The upperside resembles the light forms of the Saepiolus Blue.

(5, 6, 7) SONORA BLUE appears very early in spring. It is found near stonecrop (*Sedum*) and other succulent plants. Larva feeds in the thick plant tissues.

(2, 3, 4) SAEPIOLUS BLUE is variable; some forms in the West are dark. The row of tiny orange spots on the underside of hindwing is distinctive.

(8, 9, 10) SQUARE-SPOTTED BLUE occurs in June or July where its food, *Eriogonum*, grows. Black spots on underside are squarish. Resembles Acmon Blue.

Silvery

Sonora

Saepiolus

Square-spotted

73

SKIPPERS

Skippers (more than 3,000 kinds) are distinguished from true butterflies by the antennae, which are farther apart at the base and end in pointed, curved clubs. Skippers are named for their skipping flight. Most are drab. Many are difficult to distinguish. Their bodies are robust and moth-like. The larvae, distinctly narrowed behind the head, rest during the day between leaves pulled loosely together by silk strands. The smooth pupae are formed in similar shelters, often on the ground.

Silver-spotted Skipper
1.7-2.0"

1.5-1.8"

underside

larva

pupa

HOARY-EDGED, or Frosted, Skipper occurs from southern New England to Fla. and west to Tex. and Iowa. Underside resembles Silver-spotted. Larva feeds mostly on tick trefoils.

SILVER-SPOTTED SKIPPER, common throughout the warm seasons from so. Canada through Cen. Amer., has distinctive silver patch on the hindwing. The larva feeds on locusts and wisteria.

1.3-1.6"

1.6-2.0"

▲

NORTHERN CLOUDY WING

is difficult to tell from Southern Cloudy Wing. The spots on the forewing are usually smaller and the wing fringes are darker. It is double-brooded in the North, may have three or more broods in the South, and overwinters in the pupal stage. The green larva, which lives in a silk-lined nest, feeds on clover and other herbaceous legumes.

GOLDEN-BANDED SKIPPER

is generally uncommon and unusually sluggish. It occurs in wet woodlands from N.Y. south and west to Arizona. The larva is light green with yellow dots.

LONG TAILED SKIPPER is

also called the Bean Leaf Roller because of the way the larva attacks cultivated beans. Common in the Southeast. It overwinters as a pupa.

SOUTHERN CLOUDY WING

prefers woods margins, especially near clover and other legumes on which the brown larva feeds. Note the larger white spots on the forewings.

▼

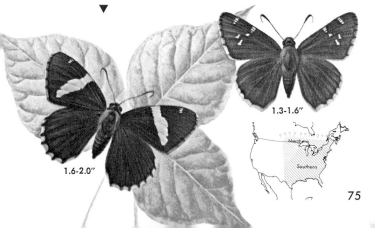

1.6-2.0"

1.3-1.6"

75

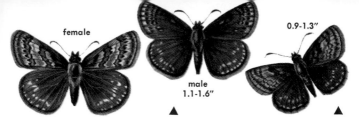

female

0.9-1.3"

male
1.1-1.6"

SLEEPY DUSKY-WING, unlike most other dusky-wings, has no clear spots on the forewing. Found from southern Canada to the Gulf and west to the Rockies.

DREAMY DUSKY-WING is smaller than Sleepy Dusky-wing; also has no clear spots on the forewing. Like the Sleepy, it occurs in early spring.

1.2-1.8"

1.0-1.3"

JUVENAL'S DUSKY-WING occurs in woods margins in Sleepy's range in spring. Hindwing has two distinct white spots below. Female is paler than male.

MOTTLED DUSKY-WING has white marks of upperside repeated below. Female is lighter than the male. Adults occur in late May and mid-July.

MOURNFUL DUSKY-WING of the West Coast has white dashes on the underside next to the white fringe on the hindwing.

FUNEREAL DUSKY-WING, found southwest from Colo. and Tex., has white fringes on the hindwings. Larva feeds on alfalfa.

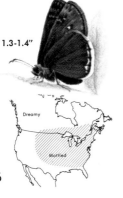

1.3-1.4"

1.3-1.8"

1.0-1.3"

1.8-2.3"

CHECKERED SKIPPER, a common species, varies greatly in the amount of gray. Flies fast without the characteristic skipping motion. Larva feeds on mallows.

GRIZZLED SKIPPER resembles Checkered Skipper, but the forewing spots are somewhat square and separate; its underside is darker. Also feeds on mallows.

BRAZILIAN SKIPPER (or Larger Canna Leaf Roller) ranges from Argentina north to Tex., So. Car., and occasionally northward. Larva feeds on opening leaves of cannas; may be destructive.

0.8-1.1"

Common Sooty-wing
with larva
and pupa
1.0-1.2"

pupa

Southern
Sooty-wing
1.0-1.3"

larva

COMMON SOOTY-WING occurs throughout North America. It overwinters as a larva on the food plant, pigweed, between leaves that have been rolled together with silk.

SOUTHERN SOOTY-WING is like the Common, but wings have faint dark bands. Occurs from Pa. to Nebr. and the S.E. Larva feeds on Lamb's Quarters.

Grizzled

Checkered

77

0.7-1.0"

underside

1.0-1.4"

▲   ▲

LEAST SKIPPER, common east of the Rockies from spring to fall, flies close to the ground, usually in marshy areas. It varies in the amount of orange above.

UNCAS SKIPPER of the Great Plains can be recognized by the dark patches around the white spots on the underside of the hindwing.

COBWEB SKIPPER occurs in early spring from Wisconsin and Texas eastward. It and the Indian Skipper resemble Uncas Skipper.

JUBA SKIPPER is found in sagebrush regions from the Pacific coastal states east to Colorado. Seen in both spring, fall.

▼   1.3-1.4"   ▼

1.0-1.3"   1.0-1.4"

INDIAN SKIPPER appears in early spring in eastern United States and Canada. Larva feeds on Panic Grass.

LEONARD'S SKIPPER, a late summer, eastern species, frequents wet meadows and open regions. The larva feeds on grass and overwinters when small.

GOLDEN SKIPPER of the arid Southwest occurs from April to September. The underside is plain yellow. Larva feeds on Bermuda Grass (*Cynodon dactylon*).

▼   ▼

1.1-1.4"   underside   0.7-1.0"

female
0.9-1.2"

male
1.0-1.3"

female

BROKEN DASH is named for the dash mark on upperside of male's forewing. Unlike Peck's, it is broken into a long upper mark and a lower dot. Both sexes lack Peck's yellow patches on undersides. Found east of Rockies.

LONG DASH has an irregular, dark, oblique mark on forewing of male. Yellow areas, less distinct than in Peck's, occur on undersides in both sexes. Long Dash occurs from Virginia and Illinois north into Canada.

0.9-1.3"

male
0.9-1.0"

female

VERNAL SKIPPER, or Little Glassy Wing, is a midsummer species found east of the Rockies. Female resembles male. The larva feeds on grass.

PECK'S SKIPPER occurs from so. Canada and New Eng. south to Fla. and west to Kans. and Ariz. Undersides of both sexes have distinct yellow patches.

FIERY SKIPPER has characteristic short antennae and varies in color pattern. The female is dark brown with a few light spots above.

FIELD SKIPPER, or Sachem, occurs in the South in spring. By midsummer it ranges to N.Y., N. Dak., and San Francisco. Larva feeds on Bermuda Grass.

1.0-1.3"

male 1.2-1.4"

female

79

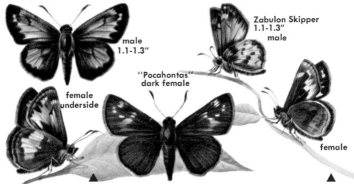

HOBOMOK SKIPPER occurs from Kansas to southern Canada. Female has two color forms, one like the male and the other, or "Pocahontas," being quite dark and different above.

ZABULON SKIPPER is found from Massachusetts to Texas. Male resembles Hobomok and female looks like the Pocahontas form of Hobomok. Zabulon and Hobomok both feed on grasses.

ROADSIDE SKIPPER has hindwings without markings. Wing fringes are strongly checkered. It occurs from southeast Canada to Florida and west to California. The pupa winters on grass.

OCOLA SKIPPER ranges as far north as N.Y., and is common from Virginia and Arkansas south to Florida and Texas. The markings of its forewings are repeated on the underside.

YUCCA SKIPPER occurs in semi-arid regions from the Carolinas to Florida and west to California. Females are larger than males, with a row of four yellow spots on the hindwing. Both have broad wings and stout bodies. Larva feeds on yucca stalks.

MOTHS

The 8,000 or so species of moths which occur in North America north of Mexico have bodies more plump and furry than those of butterflies. At rest, moths usually hold their wings flat or fold them roof-like over their backs. Their antennae, often feathery, vary in structure but usually lack the terminal club typical of butterflies. Most moths have a frenulum, a curved

PLEBEIAN SPHINX occurs throughout eastern North America. The larva, active at night, feeds on the Trumpet Vine. See pp. 82-94 for other sphinxes.

spine or group of bristles on the inner (humeral) angle of the hindwing. This helps to hold the fore and hindwings together in flight. Most moths fly at night and are attracted to lights. A few female moths do not fly at all. Larvae of moths spin silken cocoons or pupate on the ground or in underground cells.

In this section, moths are treated as commonly recognized groups which usually, but not always, include closely related species.

Below: Types of moth antennae. *Right:* Underwing Moth: underside of wings showing frenulum.

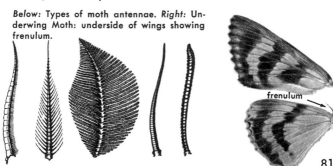

frenulum

SPHINX MOTHS, about 100 N. A. species, have large, stout larvae that hold the body erect, in a sphinx-like position. Most larvae have a horn at the rear of the body. Adults are powerful fliers; they often have a long proboscis, used to suck nectar. Some are called hawk-moths for their swooping flight; others, hummingbird moths because they hover while feeding.

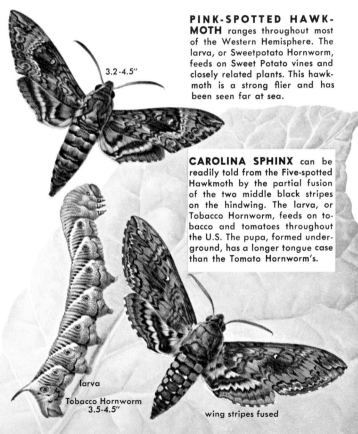

PINK-SPOTTED HAWK-MOTH ranges throughout most of the Western Hemisphere. The larva, or Sweetpotato Hornworm, feeds on Sweet Potato vines and closely related plants. This hawk-moth is a strong flier and has been seen far at sea.

3.2-4.5"

CAROLINA SPHINX can be readily told from the Five-spotted Hawkmoth by the partial fusion of the two middle black stripes on the hindwing. The larva, or Tobacco Hornworm, feeds on tobacco and tomatoes throughout the U.S. The pupa, formed underground, has a longer tongue case than the Tomato Hornworm's.

larva
Tobacco Hornworm
3.5-4.5"

wing stripes fused

RUSTIC SPHINX, or Six-spotted Sphinx, is common in southern states. On the sides of the body are three pairs of yellow spots. Larva resembles Tobacco and Tomato Hornworms, but the skin is rough instead of smooth where the markings appear. Feeds mainly on the Fringe Tree and jasmine.

4.0-5.0"

Tomato Hornworm larva

FIVE-SPOTTED HAWK-MOTH, unlike Carolina Sphinx, has the black stripes on the hindwing separated. Larva, the Tomato Hornworm, commonly feeds on tobacco. White abdominal marks form an angle instead of a single oblique line as on the Tobacco Hornworm. Widespread in temperate N.A.

pupa

wing stripes separated

3.5-4.5"

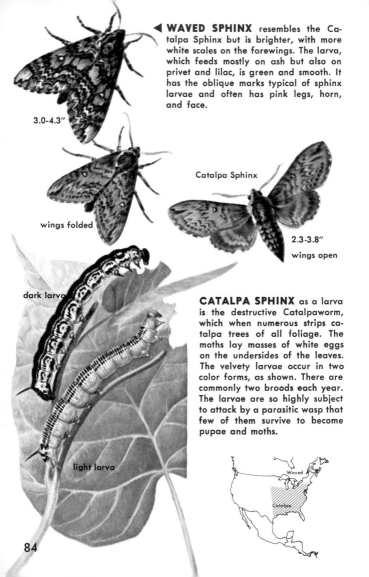

◄ WAVED SPHINX resembles the Catalpa Sphinx but is brighter, with more white scales on the forewings. The larva, which feeds mostly on ash but also on privet and lilac, is green and smooth. It has the oblique marks typical of sphinx larvae and often has pink legs, horn, and face.

3.0-4.3"

wings folded

Catalpa Sphinx

2.3-3.8"
wings open

dark larva

light larva

CATALPA SPHINX as a larva is the destructive Catalpaworm, which when numerous strips catalpa trees of all foliage. The moths lay masses of white eggs on the undersides of the leaves. The velvety larvae occur in two color forms, as shown. There are commonly two broods each year. The larvae are so highly subject to attack by a parasitic wasp that few of them survive to become pupae and moths.

Waved

Catalpa

84

FOUR-HORNED SPHINX, or Elm Sphinx, occurs from Canada to Florida, and west through the Miss. Valley. Adult has paler color on front margin of forewing than Catalpa and Waved sphinxes. The larva, which feeds on elm or birch, has four green or brown rough projections on the thoracic segments.

3.0-4.8"

4-horned Sphinx larva on elm

HERMIT SPHINX larva, which feeds on mints, has an abrupt hump near its front end. Hermitlike Sphinx, of the Southwest, is similar but is light gray and lacks the long black stripe down the middle of its abdomen. ▶

PAWPAW SPHINX is brown of varying shades, with small white spots on each side of the center abdominal stripe. The larva feeds on leaves of Pawpaw and Black Alder. ▶

2.3-3.0"

1.8-2.8"

Hermit

Pawpaw

85

3.3-4.3"

ELEGANT SPHINX is much like the Great Ash Sphinx but is a darker gray. These moths are frequently seen at evening primrose flowers. The larva is sometimes a pest of apple and plum trees.

Great Ash Sphinx larva

3.8-4.5"

GREAT ASH SPHINX, or Pen-marked Sphinx, gets the latter name from the black wavy streaks on the forewings. Its thorax is lighter than that of the Apple Sphinx or Wild Cherry Sphinx and lacks the white streaks on the sides. The larva feeds mostly on ash trees but also on lilac and privet. It is double-brooded in some areas, but in others it occurs in all stages from spring to fall.

Elegant

Great Ash

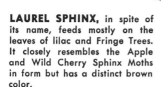

LAUREL SPHINX, in spite of its name, feeds mostly on the leaves of lilac and Fringe Trees. It closely resembles the Apple and Wild Cherry Sphinx Moths in form but has a distinct brown color.

3.3-4.3"

2.5-3.5"

APPLE SPHINX resembles the Wild Cherry Sphinx but lacks both the white shading along the front edge of the forewing and the black band down the side of the abdomen. It occurs in midsummer. The larva is bright green and has seven slanted white lines edged with pink. It feeds mainly on apple, ash, wild rose, Myrtle and Sweet Fern.

3.0-4.5"

WILD CHERRY SPHINX looks much like Apple Sphinx but has a lateral black band on the abdomen. The larva feeds on cherry, plum, and apple. Unlike most other hawkmoth larvae, it hides during the day. It is darker than the larva of Great Ash Sphinx and has violet body stripes. In the Apple Sphinx these are pinkish; in the Laurel Sphinx they are often bluish marked with black.

ELLO SPHINX is common from the Gulf states to the tropics and often strays northward. Females lack the dark streaks in the forewings. The larva feeds mostly on poinsettia and cassava.

larva

2.8-4.0"

1.8-2.3"

ABBOT'S PINE SPHINX is somewhat variable in color pattern and resembles Northern, as does the larva. The larva feeds on pines and, like the Northern larva, may become destructive.

NORTHERN PINE SPHINX and Abbot's Pine tend to intergrade where ranges overlap. The larva, with triangular head and no typical caudal horn, feeds on white, pitch, and jack pines.

1.8-2.8"

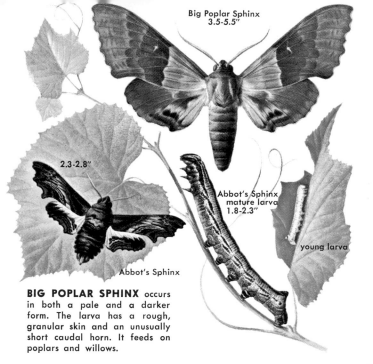

Big Poplar Sphinx
3.5-5.5"

2.3-2.8"

Abbot's Sphinx
mature larva
1.8-2.3"

young larva

Abbot's Sphinx

BIG POPLAR SPHINX occurs in both a pale and a darker form. The larva has a rough, granular skin and an unusually short caudal horn. It feeds on poplars and willows.

ABBOT'S SPHINX has two forms of mature larva, one shown above and another bright green with brown spots. Young larvae are like that illustrated. Larvae feed on grape and woodbine.

SEQUOIA SPHINX was first found resting on a Sequoia tree and was so named. It frequents the blossoms of Wild Cherry and buckeye. The larva feeds on Wild Cherry leaves.

Big Poplar

Sequoia

Abbot

1.8-2.3"

ONE-EYED SPHINX is darker and more common in the western part of its range. Larva (on willow) resembles the Small-eyed Sphinx but lacks red spots and has a pink, violet, or blue horn.

2.3-3.3"

HUCKLEBERRY SPHINX may be confused with Small-eyed Sphinx, but the outer edge of the forewing is straight instead of concave. The larva feeds on blueberry and huckleberry.

2.0-2.8"

SMALL-EYED SPHINX and other eye-spotted sphinxes rest with a lobe of the hindwing extended before the forewing and, in the case of the male, with the abdomen curved upward. Spots on larva vary or are absent. Prefers Wild Cherry but also feeds on birch and other trees. Occurs from southern Canada to Florida and west to the Rockies.

2.0-2.5"

wings open

larva

wings folded

One-eyed

Huckleberry

TWIN-SPOTTED SPHINX gets its name from the blue bar across each eyespot on the hindwing. Larva cannot be told from that of One-eyed Sphinx. It feeds on Wild Cherry, birch, and willow.

BLINDED SPHINX resembles the Small-eyed, but its forewings have scalloped margins. The larva also looks like that of Small-eyed. It feeds on a variety of trees but prefers birch, willow, and cherry.

WALNUT SPHINX is a relatively common sphinx moth found from New England south to Florida and west to Manitoba and Texas. The active larva feeds only on walnut, Butternut, pecan, and hickory, and is sometimes common in pecan orchards. The moths vary considerably in color. The larvae vary in color also, from green to reddish.

2.0-2.8"

2.3-2.8"

1.8-2.8"

larva

Twin-spotted

Blinded

1.8-2.0"

2.0-2.8"

NESSUS SPHINX flies at early dusk. The larva resembles that of the Hog Sphinx but has a shorter horn and more oblique marks on the side of the body. It feeds on grape and Virginia Creeper.

HYDRANGEA SPHINX occurs in much the same range as Hog Sphinx. At rest it assumes the position shown for Hog Sphinx. The larva feeds on Hydrangea, Buttonbush, and Swamp Loosestrife.

AZALEA SPHINX resembles the Hog Sphinx also, but the forewing is brown instead of greenish and the hindwings are entirely orange-brown. The larva feeds on viburnum and azalea.

HOG SPHINX, or Virginia-Creeper Sphinx, shown at rest, has hindwings almost entirely bright orange-brown. It is common and sometimes becomes a pest in vineyards. Unlike most hawkmoth larvae that burrow in the ground to pupate, a Hog Sphinx larva forms a loose cocoon of silk among dead leaves on the ground.

2.0-2.8"

larva

Hog Sphinx

2.0-2.8"

pupa

Nessus

Azalea and Hog

3.0-4.0"

Pandorus Spinx larva

ACHEMON SPHINX larva resembles that of Pandorus Sphinx except that the spots on the sides of the body are long and angular instead of oval. This species feeds on grape and Virginia Creeper.

Lesser Vine Sphinx

3.3-4.3"

3.0-4.5"

▲
PANDORUS SPHINX, when an almost half-grown larva, loses its horn and acquires a glassy eye-spot in its place. The larva, green or reddish brown, feeds on grape and Virginia Creeper.

LESSER VINE SPHINX, a tropical species, strays into New Eng. Larva feeds on grape and Va. Creeper. It is marked with black, white, and red, and has an eye-spot in place of a horn.

Achemon   Pandorus

93

1.5-2.5"

larva

Hummingbird Moth

Snowberry Clearwing **1.3-1.8"**

White-lined Sphinx and larva **2.0-3.5"**

Galium Sphinx

2.0-2.8"

HUMMINGBIRD MOTH, or Common Clearwing, varies in color with season and race. The "clear" wing appears as scales wear off soon after the moth emerges. Feeds at flowers by day.

SNOWBERRY CLEARWING also has seasonal and racial forms. It differs from the Hummingbird Moth in having an unscaled cell on the front edge of the forewing near the body.

WHITE-LINED SPHINX, or Striped Morning Sphinx, often flies by day. The larva is sometimes green with a series of yellow spots. When abundant it is a pest, feeding on many broad-leaved plants from southern Canada into Central America.

GALIUM SPHINX is from the Old World. The larva, like that of the White-lined Sphinx, has two color forms. In Europe it feeds on bedstraw (Galium) but in America mainly on Epilobium and other plants.

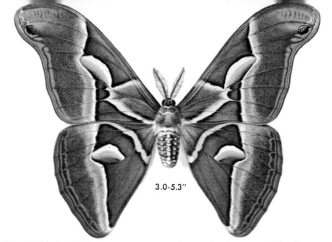

3.0-5.3"

CYNTHIA MOTH, introduced from China, is found in cities from Boston to Savannah and westward to Indiana. The larva, which feeds on Ailanthus, resembles the Cecropia larva, but all tubercles are blue. The cocoons hang like those of the Promethea Moth.

GIANT SILK MOTHS, most of which are large and attractive, number about 42 species north of Mexico. Some have clear spots, or "windows," in their wings. In some the sexes differ in size and color, but males can always be told by their more feathery antennae. The proboscis is barely developed, indicating that adults do not feed. The hindwing has no frenulum. The larva, which feeds mostly on leaves of trees and shrubs, is ornately armed with tubercles and spines. The cocoon, long and oval, is made of silk and is attached to the food plants. Easily spotted in winter. These night-flying moths come to lights, and unmated females attract distant males. You can obtain specimens and study mating, egg laying, and growth by placing a newly emerged female in an out-of-doors cage and waiting for the males to reach the cage.

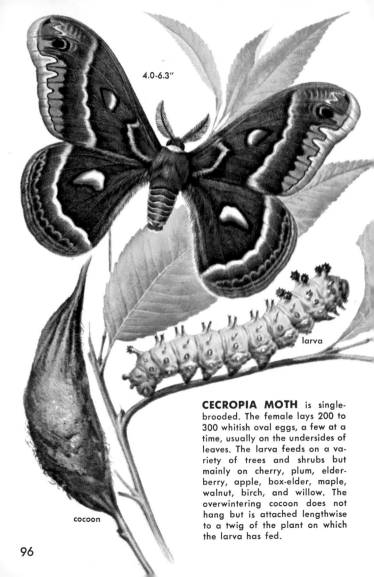

4.0-6.3"

larva

cocoon

CECROPIA MOTH is single-brooded. The female lays 200 to 300 whitish oval eggs, a few at a time, usually on the undersides of leaves. The larva feeds on a variety of trees and shrubs but mainly on cherry, plum, elderberry, apple, box-elder, maple, walnut, birch, and willow. The overwintering cocoon does not hang but is attached lengthwise to a twig of the plant on which the larva has fed.

Glover's Silk Moth
3.5-5.0"

Ceanothus Silk Moth
3.5-5.0"

GLOVER'S SILK MOTH resembles Cecropia in early stages, but mature larva has yellow instead of red thoracic tubercles. Feeds on cherry, willow, alder, wild currant, and other plants.

CEANOTHUS SILK MOTH is single-brooded. Larva resembles that of Glover's, but its yellow thoracic tubercles are ringed with black at the middle. Prefers buckbush *(Ceanothus)* leaves.

97

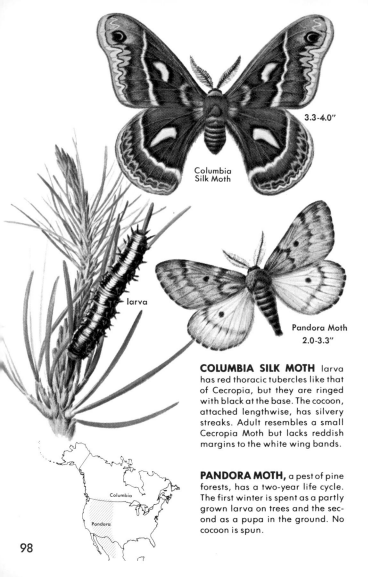

3.3-4.0"

Columbia
Silk Moth

larva

Pandora Moth
2.0-3.3"

COLUMBIA SILK MOTH larva has red thoracic tubercles like that of Cecropia, but they are ringed with black at the base. The cocoon, attached lengthwise, has silvery streaks. Adult resembles a small Cecropia Moth but lacks reddish margins to the white wing bands.

PANDORA MOTH, a pest of pine forests, has a two-year life cycle. The first winter is spent as a partly grown larva on trees and the second as a pupa in the ground. No cocoon is spun.

Columbia

Pandora

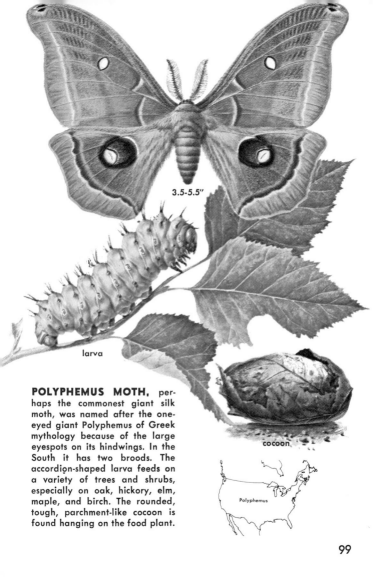

3.5-5.5"

larva

cocoon

POLYPHEMUS MOTH, perhaps the commonest giant silk moth, was named after the one-eyed giant Polyphemus of Greek mythology because of the large eyespots on its hindwings. In the South it has two broods. The accordion-shaped larva feeds on a variety of trees and shrubs, especially on oak, hickory, elm, maple, and birch. The rounded, tough, parchment-like cocoon is found hanging on the food plant.

Polyphemus

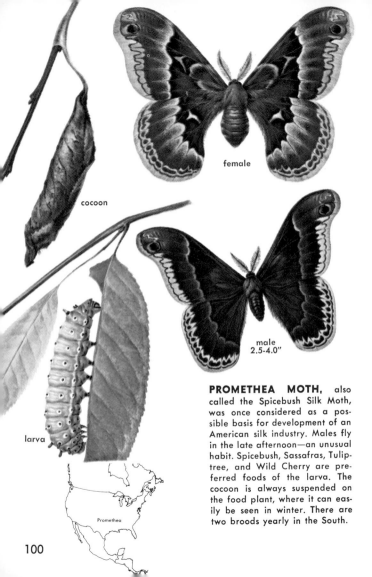

female

cocoon

male
2.5-4.0"

larva

Promethea

PROMETHEA MOTH, also called the Spicebush Silk Moth, was once considered as a possible basis for development of an American silk industry. Males fly in the late afternoon—an unusual habit. Spicebush, Sassafras, Tuliptree, and Wild Cherry are preferred foods of the larva. The cocoon is always suspended on the food plant, where it can easily be seen in winter. There are two broods yearly in the South.

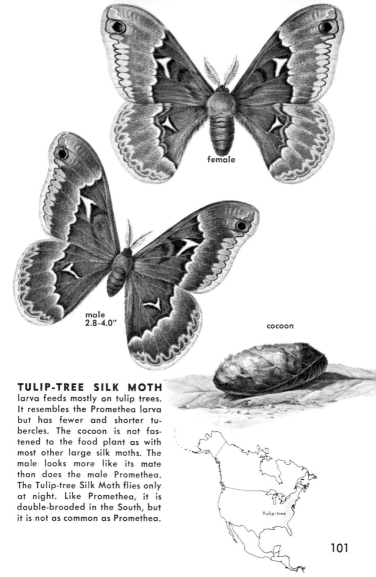

female

male
2.8-4.0"

cocoon

TULIP-TREE SILK MOTH

larva feeds mostly on tulip trees. It resembles the Promethea larva but has fewer and shorter tubercles. The cocoon is not fastened to the food plant as with most other large silk moths. The male looks more like its mate than does the male Promethea. The Tulip-tree Silk Moth flies only at night. Like Promethea, it is double-brooded in the South, but it is not as common as Promethea.

Tulip-tree

101

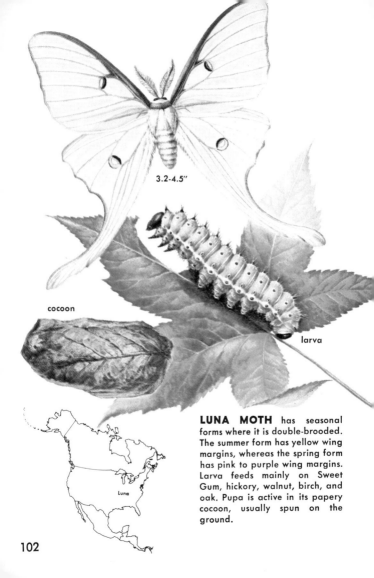

3.2-4.5"

cocoon

larva

LUNA MOTH has seasonal forms where it is double-brooded. The summer form has yellow wing margins, whereas the spring form has pink to purple wing margins. Larva feeds mainly on Sweet Gum, hickory, walnut, birch, and oak. Pupa is active in its papery cocoon, usually spun on the ground.

Luna

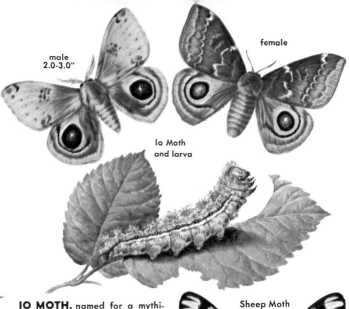

male
2.0-3.0"

female

Io Moth
and larva

IO MOTH, named for a mythical Greek maiden, has conspicuous eyespots on the hindwings. Eggs are laid in clusters. The larvae, which have irritating spines, stay together and move in long trains. They feed on a wide variety of plants, including corn and roses. The larvae spin thin, papery cocoons on the ground.

SHEEP MOTH eggs, laid in masses around twigs, hatch in the spring. The larva feeds mainly on plants of the rose family. When mature it is brownish-black with tan and black spines, red spots down the middle, and a red line on each side. Pupates early. The moth emerges in the summer, some in the second summer.

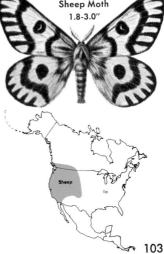

Sheep Moth
1.8-3.0"

Sheep

Io

1.5-2.8"

BUCK MOTH flies by day. In fall, females lay eggs in clusters around a twig, usually oak. The eggs hatch in spring. Like those of the Io, Buck larvae feed together and have stinging spines. Larvae do not spin cocoons but burrow into the ground and pupate. Most moths emerge in fall, some the next spring or the following fall.

larva

eggs

1.8-3.0"

NEVADA BUCK MOTH is like the Buck Moth in most respects. The larva is greenish instead of gray and has brown spots on each side of the back and along the sides of the body. The spiracles are yellow, edged with brown. The larva feeds on willow and poplar. Moths are found from September to November.

Nevada   Buck

Range
Caterpillar

RANGE CATERPILLAR, a serious range pest in the Southwest, feeds on grasses and sometimes attacks corn and other crops. Moths occur in the fall and, although day-flying, are attracted to lights at night. The eggs are laid in masses around plant stems. The cocoon is of loosely joined plant fragments.

1.5-2.5"

TRUE SILK MOTHS are not native to North America. The silkworm that produces the silk used for thread comes from Asia, where the Chinese first learned to unravel the silk from cocoons some 5,000 years ago. In commercial silk production the moths are induced to lay eggs on cards. The eggs hatch in about 10 days, and the "worms" are fed mulberry leaves. They eat steadily until in about a month the silkworm becomes full-grown. Soon every larva is ready to spin a cocoon. A few cocoons are allowed to develop into moths, but most are placed in boiling water so they can be unraveled easily. The single strand of silk that makes each cocoon may be from 500 to 1,300 yards long. Strands are combined to make a thread not yet duplicated by any synthetic fiber.

larva

cocoon

1.3-1.8"

Silk Moth

105

REGAL, OR ROYAL, MOTHS are medium- to large-sized. There are fewer than twenty species in America north of Mexico. Caterpillars, generally spiny, feed on many kinds of trees. Larvae of some species are destructive to forests and shade plantings. The caterpillars do not spin cocoons but burrow into the ground, where the pupa is formed. At rest, Regal Moths usually fold their wings roof-like over their bodies. Like silk moths they are not attracted to bait but are attracted to lights.

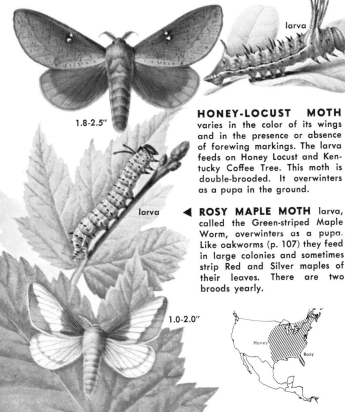

larva

1.8-2.5"

larva

HONEY-LOCUST MOTH varies in the color of its wings and in the presence or absence of forewing markings. The larva feeds on Honey Locust and Kentucky Coffee Tree. This moth is double-brooded. It overwinters as a pupa in the ground.

◄ **ROSY MAPLE MOTH** larva, called the Green-striped Maple Worm, overwinters as a pupa. Like oakworms (p. 107) they feed in large colonies and sometimes strip Red and Silver maples of their leaves. There are two broods yearly.

1.0-2.0"

Honey

Rosy

OAKWORM MOTHS are easier to tell apart as larvae than as adults. Larvae feed in colonies, sometimes so populous as to completely strip forests. Females are larger than males, with thinner antennae and stouter bodies.

ORANGE-STRIPED OAK-WORM The male has angular shaped hindwings. The forewings of the female are thinner and less speckled than those of the female Spiny Oakworm.

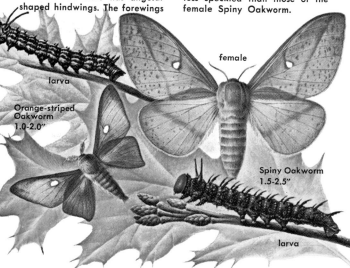

larva

Orange-striped
Oakworm
1.0-2.0"

female

Spiny Oakworm
1.5-2.5"

larva

SPINY OAKWORM is the largest of the species illustrated. The male resembles the female more closely than in the other Oakworm species.

PINK-STRIPED OAKWORM Male has narrower, more triangular forewings, thinner beyond the spot than other species. Female lacks spotting on wings.

Spiny, Orange-striped
and Pink-striped

1.5-2.3"

107

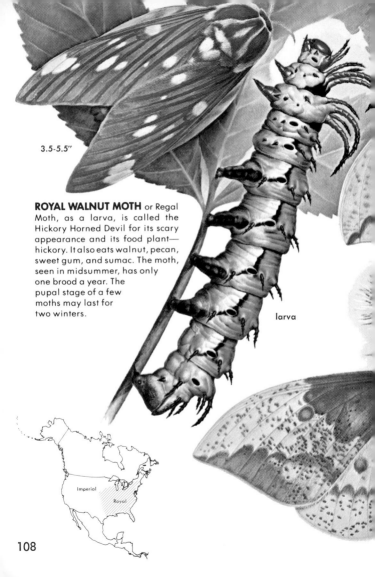

ROYAL WALNUT MOTH or Regal Moth, as a larva, is called the Hickory Horned Devil for its scary appearance and its food plant—hickory. It also eats walnut, pecan, sweet gum, and sumac. The moth, seen in midsummer, has only one brood a year. The pupal stage of a few moths may last for two winters.

3.5-5.5"

larva

Imperial

Royal

108

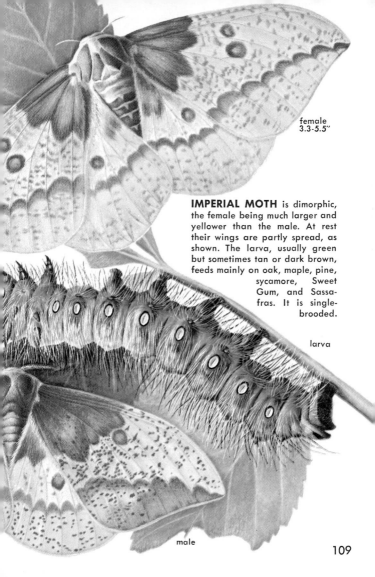

female
3.3-5.5"

IMPERIAL MOTH is dimorphic, the female being much larger and yellower than the male. At rest their wings are partly spread, as shown. The larva, usually green but sometimes tan or dark brown, feeds mainly on oak, maple, pine, sycamore, Sweet Gum, and Sassafras. It is single-brooded.

larva

male

109

TIGER MOTHS are small to medium in size and generally light in color. Many have conspicuous spots or stripes. Only a few of some 200 species north of Mexico have functional mouthparts. Adults, especially males, come readily to lights. Most larvae are covered with a dense coat of hairs, which are shed and mixed with silk when the cocoon is made. Most caterpillars move about rapidly and are active by day. They commonly roll into a ball when disturbed.

GREAT LEOPARD MOTH is well named for its wing spots. The hairs on the caterpillar are very stiff, and crimson rings between segments show distinctly when the larva rolls into a ball. Feeds on plantain. Cocoon is spun in spring. Larva overwinters.

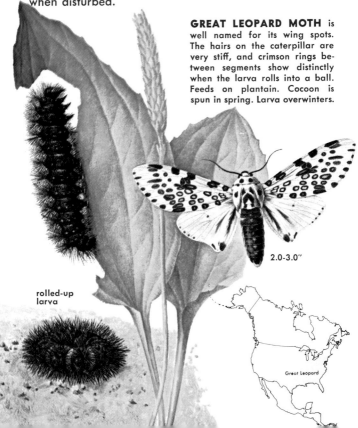

2.0-3.0"

rolled-up larva

Great Leopard

ACREA MOTH larva, called the Salt Marsh Caterpillar, feeds on herbaceous plants. It is double-brooded. The larvae are often abundant in fall and over-winter as pupae in cocoons. Adult wings are profusely spotted. The female has white hindwings.

1.8-2.5"

ISABELLA MOTH is not as well known as its caterpillar, the Banded Woolly Bear. The amount of black on each end of its body does not, of course, predict the coldness of the coming winter. There are two broods yearly; cocoons are formed in spring and summer.

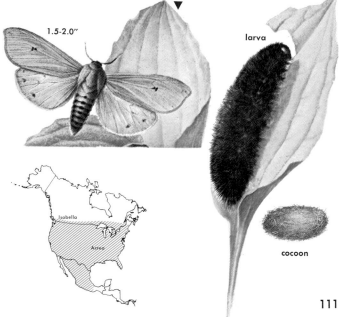

1.5-2.0"

larva

cocoon

Fall Webworm
0.8-1.5"

spotless

1.8-2.8"

larva

lightly spotted

heavily spotted

GARDEN TIGER MOTH is so variable that its forms look like different species. The hindwings may be yellow with dark spots, and the forewings may bear very broad white bands.

FALL WEBWORM is also variable. Some moths have heavily spotted wings. Others have only a few black dots. The bodies of some are yellow with black dots on the sides, while others are plain white. The social larvae extend their webs over the foliage of the many deciduous trees on which they feed. The webs may soon cover large branches. These webs are sometimes confused with those of tent caterpillars (p. 138). Large numbers of eggs are laid in masses, usually on the undersides of leaves. Pupae overwinter in cocoons.

Garden

Fall

1.5-2.0"

Arge Tiger Moth
larva

▲
CLYMENE is a Haploa Tiger Moth, of which there are five species north of Mexico. In some the hindwings are white or buff. Clymene has one brood. The moths occur in midsummer.

APANTESIS TIGER MOTHS include some 30 species, often varying greatly in color. In general their forewings are jet black with irregular white streaks, and their hindwings are yellow or red with black spots or bands. The hairy caterpillars live over the winter and feed mostly on herbaceous plants. Males are commonly captured at lights.

ARGE is double-brooded, with moths occurring June and Sept.

VIRGO has a typical Apantesis color pattern. It is single-brooded, with moths on the wing mainly in July.

Arge Tiger Moth
1.3-2.0"

Virgo Tiger Moth
1.8-3.0"

Virgo

Arge

Clymene

113

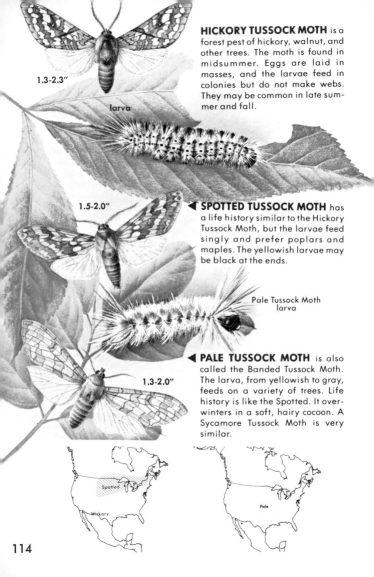

HICKORY TUSSOCK MOTH is a forest pest of hickory, walnut, and other trees. The moth is found in midsummer. Eggs are laid in masses, and the larvae feed in colonies but do not make webs. They may be common in late summer and fall.

1.3-2.3"

larva

1.5-2.0"

◀ SPOTTED TUSSOCK MOTH has a life history similar to the Hickory Tussock Moth, but the larvae feed singly and prefer poplars and maples. The yellowish larvae may be black at the ends.

Pale Tussock Moth larva

◀ PALE TUSSOCK MOTH is also called the Banded Tussock Moth. The larva, from yellowish to gray, feeds on a variety of trees. Life history is like the Spotted. It overwinters in a soft, hairy cocoon. A Sycamore Tussock Moth is very similar.

1.3-2.0"

Spotted

Hickory

Pale

114

YELLOW WOOLLY BEAR

larvae are seen more commonly than the adults. They vary from pale yellow to reddish and are confused with Acrea Moth larvae (p. 111), but have black heads. There are two broods.

larva

RANCHMAN'S TIGER MOTH ▶

is one of the largest and most attractive members of this group. Another common form of this moth has yellow hindwings and a yellow body.

1.0-1.8"

2.0-3.0"

Dogbane Tiger Moth
larva

DOGBANE TIGER MOTH ▶

varies in the amount of yellow on the front edge of the forewing. It is probably double-brooded. The caterpillar, with an unusually light hairlike covering, feeds almost entirely on dogbane. May be very common.

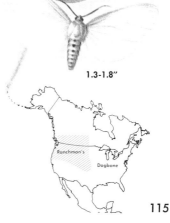

1.3-1.8"

Yellow
Woolly
Bear

Ranchman's

Dogbane

115

Showy Holomelina
1.0-1.3"

Milkweed
Tussock
larva

1.0-1.8"

1.3-1.8"

larva

pupa

SHOWY HOLOMELINA is the brightest of the eight or so holomelina moths of North America. Others are mainly yellowish, gray, or dull red. The larva, like that of the Dogbane Moth, has a soft, hairlike covering. The cocoon is very thin, with few hairs.

◄ **MILKWEED TUSSOCK MOTH** is better known by its larva, which is found in late summer feeding in colonies on milkweed. When disturbed it frequently rolls into a ball and drops from the plant. The pupa overwinters in a cocoon which is very hairy.

BELLA MOTH, or Rattlebox Moth, is locally common but is restricted to the vicinity of its food plant, Rattlebox, and other kinds of *Crotalaria*. The larva lacks the hairs typical of tiger moths. The pupa, which is unusually ornate, overwinters.

Milkweed

Showy

Bella

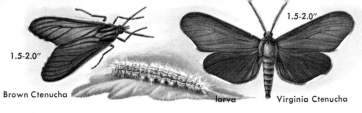

1.5-2.0"

1.5-2.0"

Brown Ctenucha   larva   Virginia Ctenucha

CTENUCHA MOTHS, daytime fliers, look like wasps when feeding at flowers. Larvae feed on marsh grasses. Hairy cocoons and larvae resemble those of some closely related tiger moths. Amounts of yellow and black hairs vary in Virginia Ctenucha larvae. Smaller, narrower-winged Brown Ctenucha Moth has a more southern range.

FORESTER MOTHS More than two dozen American species are known. They differ from most noctuid moths (p. 118) in having the ends of the antennae thickened. The Eight-spotted Forester is below. Its double-brooded larva feeds on grape and woodbine.

Eight-spotted Forester

1.0-1.5"   larva

DIOPTID MOTHS are represented by only one species in North America north of Mexico, the California Oakworm Moth. It is a pest of live oaks in California, sometimes stripping these trees.

The California Oakworm, brown with nearly transparent wings, is double-brooded. It overwinters as eggs or tiny larvae. The female California Oakworm Moth lacks the yellowish patch near the mid-forewing.

1.0-1.4"

California Oakworm larva

NOCTUID MOTHS (pp. 118-131) number over 2,600 kinds in America north of Mexico. Many are serious pests of farm and garden crops and forest and shade trees. Among these are the well-known armyworms, cutworms, and Corn Earworm. Noctuids vary greatly in habits, but most adults fly at night and feed on the nectar of flowers. Most noctuids are attracted to lights and to baits containing sugar. Some overwinter as pupae in the ground or in thin cocoons above ground. Others overwinter as larvae and a few as eggs or adults. They are sometimes called owlet moths, from the way their eyes shine in the dark when a light strikes them.

BLACK WITCH, or Giant Noctuid, is the largest of the family within our range. It is a tropical species, sometimes straying northward into Canada in the fall. Its shade of brown varies. The wings of the females come to a less sharp point at the apex. Larva feeds on acacia and similar plants. Considered a pest in Hawaii.

3.5-6.0"

AMERICAN DAGGER MOTH
is the largest of the daggers. The
larva feeds on a variety of trees.
Winter is passed as a pupa, often
in an old stump. The cocoon includes the hair of the larva.

2.0-2.8"

larva

DAGGER MOTHS (some 70 kinds) are so named for
the dagger-like mark near the forewing outer margin.
Adults are similar but many of the larvae are quite different. Some are hairy, with characteristic tufts of longer
hair called pencils. Other larvae lack hair and may be
spiny. Dagger moths overwinter as pupae.

1.3-1.8"

larva

1.3-1.8"

SMEARED DAGGER MOTH
occurs more commonly in wet-
lands, where the larva feeds on
willow, smartweed, alder, Buttonbush, and cattail. Overwintering
cocoon is thin but strong.

**COTTONWOOD DAGGER
MOTH** varies greatly in markings. Larva has soft yellow hair
like American Dagger but has
five hair pencils on its abdomen.
It feeds on poplars and willows.

119

CUTWORM MOTHS are noctuids whose larvae, or cutworms, cut off young plants just above the ground. Some cutworms drag cut-off portions of plants into a hole. Some are climbers, feeding on foliage of bushes and trees. All feed at night. Certain species occur only in spring, others only in fall; many have several broods each year. Some full-grown larvae remain at rest without feeding from spring to late summer, when they pupate.

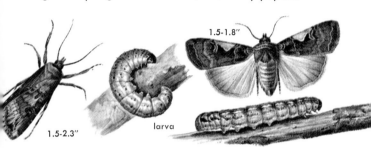

larva

1.5-1.8"

1.5-2.3"

BLACK CUTWORM, also called the Greasy Cutworm, occurs throughout the U.S. and southern Canada. Larva is a burrower, found mostly in low spots. Overwinters as a pupa.

SPOTTED CUTWORM is one of the most damaging, feeding on a wide variety of plants and often climbing them. It overwinters as a nearly mature larva. May have three broods yearly.

W-MARKED CUTWORM is a climber, feeding on a wide variety of trees, shrubs, and herbaceous plants. It overwinters as a larva and, in most areas, has two broods each year.

PALE-SIDED CUTWORM lives mostly in a tunnel into which it drags pieces of cut-off plants for food. There may be four broods yearly. The winter is passed in the pupal stage.

1.5-1.8"

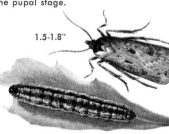

1.5-1.8"

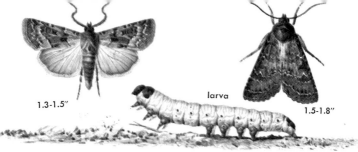

larva

1.3-1.5"

1.5-1.8"

PALE WESTERN CUTWORM
moths occur in Aug. and Sept.
and lay eggs on newly cultivated
land. Overwintering may be as
eggs or young larvae. Plants are
attacked below the soil surface.

GLASSY CUTWORM occurs
from Canada to N.J. and west
to the Pacific. It feeds on roots
and lower stems of grasses. It is
single-brooded and overwinters
as partly grown larva.

1.3-1.5"   larva   larva   1.3-1.5"

**STRIPED GARDEN CUT-
WORM** is single-brooded. The
moth occurs throughout most of
U.S. in summer. Larvae rest ex-
posed on food plants. They mature and pupate in fall.

SPOTTED-SIDED CUTWORM
has one generation. The moth
occurs from Alberta and Texas
eastward in the fall. Full-grown
larvae appear on dock and chick-
weed in early spring.

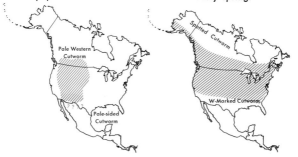

Pale Western Cutworm

Pale-sided Cutworm

Spotted Cutworm

W-Marked Cutworm

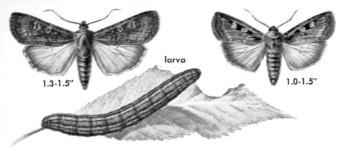

larva

1.3-1.5"

1.0-1.5"

DARK-SIDED CUTWORM is a climber, sometimes very destructive to orchards and shrubs. Eggs hatch during winter and larvae become full-grown by June. There is one brood.

STRIPED CUTWORM closely resembles Dark-sided Cutworm but differs much in life history. Eggs hatch in the fall and partly grown larvae hibernate in the soil, feeding again in spring.

1.3-2.0"

larva

1.3-1.5"

BRONZED CUTWORM prefers grasses and cereals. Moths occur in Sept. and Oct. and eggs hatch during winter. By April or May the larvae are full-grown, but they do not pupate until July.

WHITE CUTWORM sometimes damages the buds and young leaves of grapes and fruit trees. Its life history is like that of the Striped Cutworm. Head and spiracles of larva vary in color.

Dark-sided

White

Striped

122

1.1-1.5"

0.8-1.0"

DINGY CUTWORM winters as immature larva, maturing in late spring. It lies inactive (estivates) until August, when it pupates, emerging as a moth in a month or so.

BRISTLY CUTWORM often occurs in clover with the Dingy Cutworm. It is double-brooded. Larva completes growth in the spring like that of Dingy Cutworm but does not estivate.

1.0-1.8"

1.5-2.0"

VARIEGATED CUTWORM, probably the most destructive cutworm, attacks many different crops. It overwinters as a pupa and may have four broods. Occurs throughout North America.

YELLOW-STRIPED ARMY-WORM, also called Cotton Cutworm because it bores into cotton bolls, is common in the South. Feeds on many plants. Winter is passed as a pupa.

123

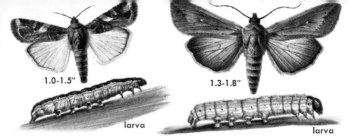

1.0-1.5"

larva

1.3-1.8"

larva

FALL ARMYWORM by mid-summer spreads north from the Gulf states but dies out by winter. In the South there may be three broods. Prefers grasses and often attacks corn.

ARMYWORM has two or more broods. The spring one is most destructive, especially to oats and small grains. Natural enemies reduce later broods. It overwinters as moth, pupa, or larva.

ARMYWORM MOTHS are noctuid moths whose larvae tend to migrate in "armies" to new feeding areas after destroying vegetation in fields where their eggs were laid. Active at night and hiding by day, the larvae feed mostly on grasses and small grains.

BEET ARMYWORM is a common pest of sugar beets in the West. Larva resembles that of Fall Armyworm, but the pale central line along the back is less distinct. Overwinters as pupa.

WHEAT HEAD ARMYWORM feeds mainly on timothy and wheat heads, attacking at night. Full-grown larva is about one inch long with narrow pale stripes. Overwinters as a pupa.

0.8-1.3"

1.0-1.5"

124

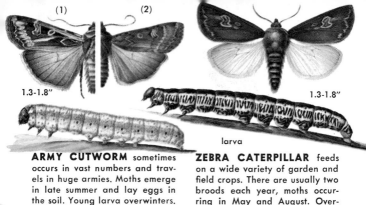

(1) **(2)**

1.3-1.8"

1.3-1.8"

larva

ARMY CUTWORM sometimes occurs in vast numbers and travels in huge armies. Moths emerge in late summer and lay eggs in the soil. Young larva overwinters. Two adult forms shown: (1), (2).

ZEBRA CATERPILLAR feeds on a wide variety of garden and field crops. There are usually two broods each year, moths occurring in May and August. Overwinters in pupal stage.

1.0-1.5"

0.8-1.5"

COTTON LEAFWORM is a slender, looping larva that feeds only on cotton. Spines at end of the moth's tongue sometimes injure ripe fruit. This tropical species cannot survive U.S. winters.

GREEN CLOVERWORM, a looping larva, feeds on clover, alfalfa, and other legumes. Winter is usually passed as an adult but some may winter as pupae. There are two to four broods.

125

1.0-1.8"

1.0-1.5"

ALFALFA LOOPER, in spite of its name, feeds on a wide variety of plants, including cereals. Winter is passed in pupal and adult stages. There are two broods, the second in July.

CABBAGE LOOPER, a common species, feeds mostly on cabbage and other members of the cabbage family. Two broods occur in the North. Hibernates as pupa in loosely woven cocoon.

LOOPERS is a name most commonly used for larvae of geometer moths (p. 140), but larvae of some noctuid moths are also called loopers because they hump their backs when crawling. The noctuid larvae have fewer than the normal four pairs of prolegs and claspers.

CELERY LOOPER in the larval stage closely resembles the Cabbage Looper. There are at least two broods. The summer form is brown, as illustrated. A spring form is gray.

BILOBED LOOPER feeds on alfalfa, clover, and many other plants. Larva resembles Cabbage Looper's but has stripes on the sides of its head. It hibernates as a pupa in a thin cocoon.

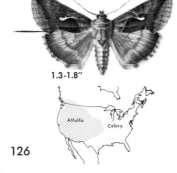

1.3-1.8"

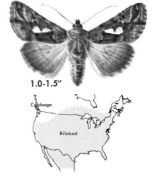

1.0-1.5"

1.5-2.0"

1.3-1.8"

FORAGE LOOPER is sometimes destructive to clover. Markings of forewings are much less distinct in females. Three broods occur from spring to fall. Pupa overwinters on leaves.

GREEN FRUITWORM eats into young apples, pears, cherries, and other fruit in spring. Moths emerge in the fall and overwinter. The larva resembles that of Copper Underwing (p. 130).

LUNATE ZALE often has large green patches on the wing margins. Pupa overwinters in soil. The larva varies greatly in color and looks like those of the underwing moths (p. 128).

DRIED LEAF MOTH, or Litter Moth, larva feeds on lichens and dead leaves. The development from tiny eggs is very slow. The moths are on the wing in midsummer and are single-brooded.

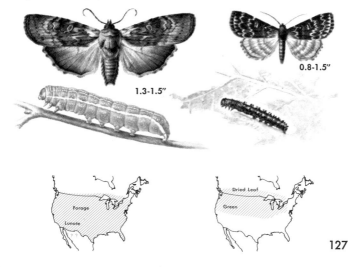

1.3-1.5"

0.8-1.5"

Forage

Lunate

Dried Leaf

Green

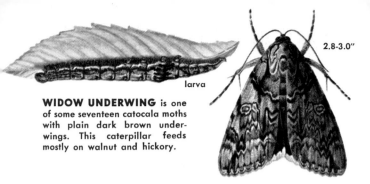

larva

WIDOW UNDERWING is one of some seventeen catocala moths with plain dark brown underwings. This caterpillar feeds mostly on walnut and hickory.

2.8-3.0"

UNDERWING, OR CATOCALA, MOTHS (about 100 kinds in America north of Mexico) readily come to light and bait. Otherwise they would rarely be seen, for in daylight they rest well camouflaged on tree trunks with their underwings hidden. Larvae also rest by day on trunks or limbs, or under debris on the ground where the thin cocoons are also found. They have one brood and overwinter as eggs on bark of trees.

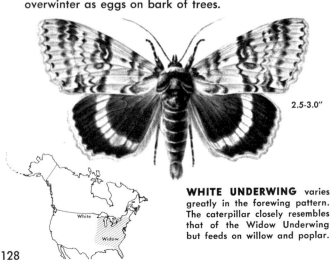

2.5-3.0"

White

Widow

WHITE UNDERWING varies greatly in the forewing pattern. The caterpillar closely resembles that of the Widow Underwing but feeds on willow and poplar.

2.5-3.3"

AHOLIBAH UNDERWING is hard to tell from several other catocala moths. Note its large size, with a wingspread occasionally exceeding three inches. Larva feeds on oak. It lacks the striped saddle patch of the Widow Underwing larva.

TINY NYMPH UNDERWING is one of several species of small catocalas with similar yellow underwings. The forewings vary greatly in pattern. The larva feeds on oak.

PENITENT UNDERWING flies from July to October. Larva lacks the fringe of hairs and swollen saddle patch of the Widow Underwing caterpillar larva. Feeds on walnut and hickory.

▼ ▼

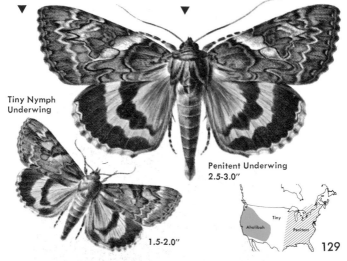

Tiny Nymph
Underwing

Penitent Underwing
2.5-3.0"

1.5-2.0"

129

2.0-2.8"

larva

LOCUST UNDERWING is not a catocala moth, despite its name, but closely resembles one. The larva feeds on Black Locust. This species is double-brooded. It overwinters in the pupal stage.

larva

1.0-1.5

larva

1.5-2.0"

larva

COPPER UNDERWING occurs from midsummer to fall but hides in cracks during the day. In the spring, larva feeds on many plants, including Woodbine. It, too, is not a catocala underwing.

PEARLY WOOD NYMPH is one of three wood nymphs. Beautiful Wood Nymph is larger and has a dark margin on the hindwing. Calif. Wood Nymph has a black dot on hindwing.

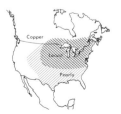

Copper

Locust

Pearly

Corn

White

Stalk

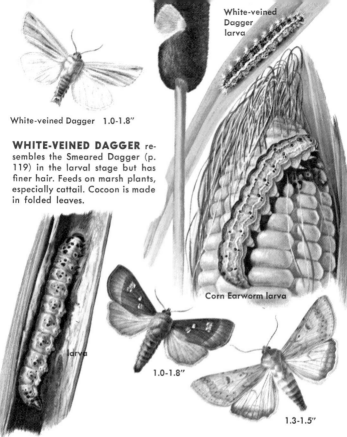

White-veined Dagger larva

White-veined Dagger   1.0-1.8"

WHITE-VEINED DAGGER resembles the Smeared Dagger (p. 119) in the larval stage but has finer hair. Feeds on marsh plants, especially cattail. Cocoon is made in folded leaves.

Corn Earworm larva

larva

1.0-1.8"

1.3-1.5"

STALK BORER, a pest of corn, feeds in the stalks of many plants, especially Giant Ragweed. Eggs laid in fall hatch very early in spring. Bright body stripes are lost as the larva matures in summer. There is only one generation annually.

CORN EARWORM, the familiar worm in ears of corn, is also called Tomato Fruitworm and Bollworm. Larvae are often found in the fruit of many plants. Winter is passed as a pupa in the soil. There may be several broods each year.

131

THE PROMINENTS, numbering about 100 species north of Mexico, resemble noctuids. Readily attracted to lights. Many of them can be told by their hairy legs when at rest. Larvae of most species live on tree leaves. Many species lack anal prolegs and hold their rear ends erect.

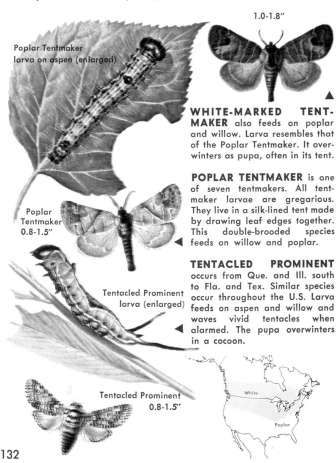

1.0-1.8"

Poplar Tentmaker
larva on aspen (enlarged)

Poplar
Tentmaker
0.8-1.5"

Tentacled Prominent
larva (enlarged)

Tentacled Prominent
0.8-1.5"

WHITE-MARKED TENT-MAKER also feeds on poplar and willow. Larva resembles that of the Poplar Tentmaker. It overwinters as pupa, often in its tent.

POPLAR TENTMAKER is one of seven tentmakers. All tentmaker larvae are gregarious. They live in a silk-lined tent made by drawing leaf edges together. This double-brooded species feeds on willow and poplar.

TENTACLED PROMINENT occurs from Que. and Ill. south to Fla. and Tex. Similar species occur throughout the U.S. Larva feeds on aspen and willow and waves vivid tentacles when alarmed. The pupa overwinters in a cocoon.

White

Poplar

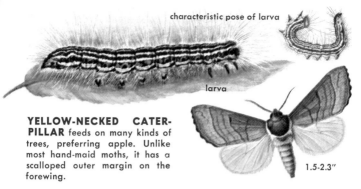

characteristic pose of larva

larva

YELLOW-NECKED CATER-PILLAR feeds on many kinds of trees, preferring apple. Unlike most hand-maid moths, it has a scalloped outer margin on the forewing.

1.5-2.3"

HAND-MAID MOTHS number 12 species that resemble one another closely. The larvae feed in colonies and may be numerous enough to strip trees. When disturbed they hold both ends of their bodies erect. Most species are single-brooded, overwintering as pupae in the ground. The Walnut Caterpillar feeds on walnut and hickory. The Sumac Caterpillar feeds only on sumac. The outer margins of the forewings of these two moths are straight. The moth of the Walnut Caterpillar is dark.

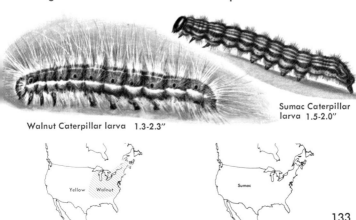

Sumac Caterpillar larva 1.5-2.0"

Walnut Caterpillar larva 1.3-2.3"

Yellow   Walnut

Sumac

133

larva

Variable
Oak Leaf 1.3-1.8"

larva

SADDLED PROMINENT is so named for the saddle mark on the back of the green, pointed-tailed larva, which feeds mainly on leaves of beech and maple. The moth resembles that of the Variable Oak Leaf Caterpillar.

VARIABLE OAK LEAF CATERPILLAR, found in late summer and fall, may lay 500 eggs singly, on oak. Larva overwinters in the soil and pupates in spring.

larva

1.3-1.5"

ANGUINA MOTH is commonly recognized by its unusual caterpillar, which feeds on various legumes, especially locust. The Anguina Moth is double-brooded.

larva

ELM LEAF CATERPILLAR is well camouflaged on elm foliage. Moths occur in June and August. Pupae spend winter in cocoons on the ground.

1.3-1.5"

134

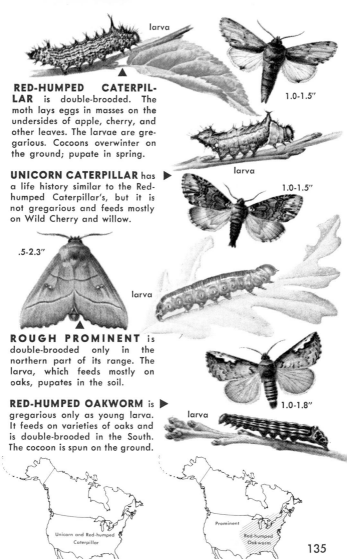

larva

RED-HUMPED CATERPIL-LAR is double-brooded. The moth lays eggs in masses on the undersides of apple, cherry, and other leaves. The larvae are gregarious. Cocoons overwinter on the ground; pupate in spring.

1.0-1.5"

UNICORN CATERPILLAR has a life history similar to the Red-humped Caterpillar's, but it is not gregarious and feeds mostly on Wild Cherry and willow.

larva

1.0-1.5"

.5-2.3"

larva

ROUGH PROMINENT is double-brooded only in the northern part of its range. The larva, which feeds mostly on oaks, pupates in the soil.

RED-HUMPED OAKWORM is gregarious only as young larva. It feeds on varieties of oaks and is double-brooded in the South. The cocoon is spun on the ground.

1.0-1.8"

larva

Unicorn and Red-humped Caterpillar

Prominent

Red-humped Oakworm

135

1.3-2.3"

pupa

larva

SATIN MOTH is Eurasian. Found in Mass. and Br. Columbia in 1920, it has spread through N.E. and N.W. U.S. and Canada. In July eggs are laid in masses. Larva prefers poplar and willow. It hibernates when partly grown.

TUSSOCK MOTHS, numbering about 30 species, get their name from the brightly colored tufts of hair on the larvae. Hairs of some species are irritating. Adult legs are hairy. Some females are almost wingless. The antennae of males are feathery. Tussock moths have no tongue. Several tussock moths, including the Satin and Gypsy Moths, are pests of forest and shade trees.

female

pupa

shed skin

larva

egg mass

male
1.0-2.5"

GYPSY MOTH, accidentally introduced from Europe about 1868 into Mass., now occurs from Can. to N.C. w. to Mich. and Ill. It overwinters as eggs. Larvae feed gregariously on many kinds of trees, especially oaks. They become full grown in July.

136

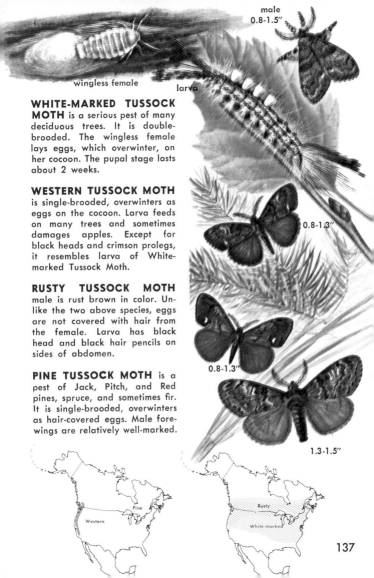

male
0.8-1.5"

wingless female

larva

WHITE-MARKED TUSSOCK MOTH
is a serious pest of many deciduous trees. It is double-brooded. The wingless female lays eggs, which overwinter, on her cocoon. The pupal stage lasts about 2 weeks.

WESTERN TUSSOCK MOTH
is single-brooded, overwinters as eggs on the cocoon. Larva feeds on many trees and sometimes damages apples. Except for black heads and crimson prolegs, it resembles larva of White-marked Tussock Moth.

0.8-1.3"

RUSTY TUSSOCK MOTH
male is rust brown in color. Unlike the two above species, eggs are not covered with hair from the female. Larva has black head and black hair pencils on sides of abdomen.

0.8-1.3"

PINE TUSSOCK MOTH
is a pest of Jack, Pitch, and Red pines, spruce, and sometimes fir. It is single-brooded, overwinters as hair-covered eggs. Male forewings are relatively well-marked.

1.3-1.5"

Pine

Western

Rusty

White-marked

137

LASIOCAMPIDS are a family of some thirty North American moths of medium size with stout hairy bodies. They are readily attracted to lights. Females are like the males but larger. The most familiar species are the tent caterpillars, pests of forest, shade, and orchard trees.

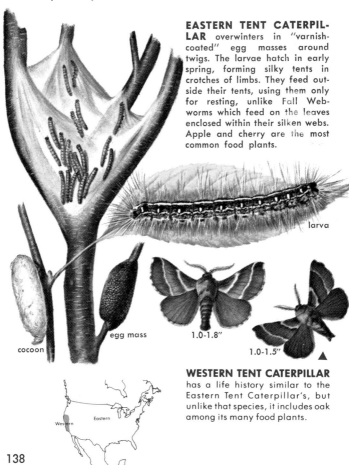

EASTERN TENT CATERPILLAR overwinters in "varnish-coated" egg masses around twigs. The larvae hatch in early spring, forming silky tents in crotches of limbs. They feed outside their tents, using them only for resting, unlike Fall Webworms which feed on the leaves enclosed within their silken webs. Apple and cherry are the most common food plants.

larva

cocoon

egg mass

1.0-1.8"

1.0-1.5"

WESTERN TENT CATERPILLAR has a life history similar to the Eastern Tent Caterpillar's, but unlike that species, it includes oak among its many food plants.

Western

Eastern

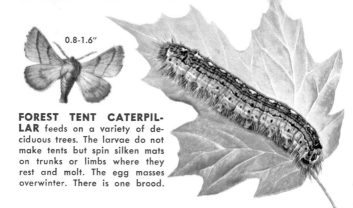

FOREST TENT CATERPILLAR feeds on a variety of deciduous trees. The larvae do not make tents but spin silken mats on trunks or limbs where they rest and molt. The egg masses overwinter. There is one brood.

0.8-1.6"

ZANOLIDS are closely related to the prominents (pp. 132-135). There are only three species in North America, all occurring east of the Rockies. Eggs are flat and wafer-like. The densely hairy larvae feed singly on leaves of trees and bushes. They make no cocoon but pupate in the ground and overwinter there.

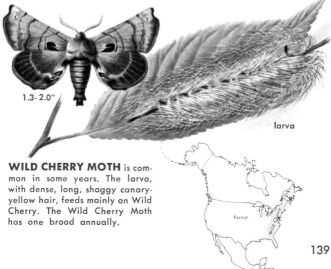

1.3-2.0"

larva

WILD CHERRY MOTH is common in some years. The larva, with dense, long, shaggy canary-yellow hair, feeds mainly on Wild Cherry. The Wild Cherry Moth has one brood annually.

Forest

GEOMETER, meaning earth measurer, refers to the way the crawling larvae of these species draw the rear of the body up to the front legs, forming a loop, and then extend the body again. The crawling pattern is associated with only two or three pairs of abdominal legs. Geometers are also called loopers, inchworms, measuring worms, and spanworms. Most geometer moths have thin bodies and relatively broad, delicate wings. The females of some species are wingless. More than 1,000 species of geometers live in North America; some are serious pests.

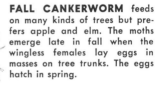

female and eggs

larva

male
0.9-1.3"

Spring
Cankerworm
0.9-1.4"

larva

FALL CANKERWORM feeds on many kinds of trees but prefers apple and elm. The moths emerge late in fall when the wingless females lay eggs in masses on tree trunks. The eggs hatch in spring.

SPRING CANKERWORM larvae differ from Fall Cankerworm's in having two rather than three abdominal prolegs, but, like them, pupate in the soil. The moths occur mostly in spring.

BRUCE SPANWORM occurs from N.J. to Quebec, and west to Alberta. The moth occurs in fall on Sugar Maple, poplar, beech, and other trees. The female is wingless. The larva resembles Spring Cankerworm's in form but has six narrow white stripes.

0.9-1.4"

140

SPEAR-MARKED BLACK-MOTH has variable amounts of black and white. It has two broods per year. Larvae live together in nests of leaves, mostly birch and willow. Pupae overwinter in soil.

light
males
dark

0.9-1.4"

Cherry Scallop-shell Moth larva

CHERRY SCALLOP-SHELL MOTH is commonly seen in the larval stage in nests of wild cherry leaves. Double-brooded. The pupa overwinters in the soil.

0.8-1.4"

CURRANT SPANWORM is a pest of currant and gooseberry. Eggs are laid in early summer but do not hatch until the next spring. Larva pupates in the soil in May or early June.

WALNUT SPANWORM moths occur in early spring. The females are wingless. Larva feeds mainly on Live Oak but also attacks English Walnut and other fruit trees. Single-brooded.

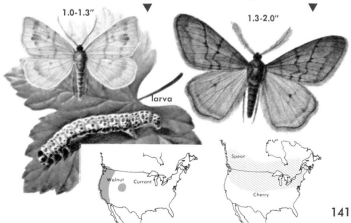

1.0-1.3"

1.3-2.0"

larva

Walnut   Currant

Spear

Cherry

141

1.4-1.8"

LINDEN LOOPER moths occur in fall. The female is wingless. Eggs are laid singly or in small groups. They hatch in spring. Larva feeds on many kinds of leaves besides linden.

larva

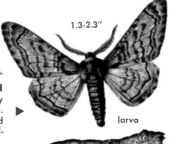

1.3-2.3"

larva

PEPPER-AND-SALT MOTH occurs in late spring and early summer. Both sexes are winged. The larva, called Cleft-headed Spanworm, feeds on many deciduous trees. Pupa overwinters.

1.1-1.4"

larva

ELM SPANWORM as an adult is called Snow-white Linden Moth. Groups of eggs are laid in July on bark of trees, pass the winter there, and hatch in spring. Larva feeds on many trees besides elm.

HEMLOCK LOOPER feeds mostly on hemlock and Balsam Fir. Moths occur in late summer. Their eggs, which hatch the following spring, are laid singly or in small groups on bark or needles.

1.1-1.5"

Linden and Elm

Pepper

Hemlock

CHAIN-SPOTTED GEOM- ETER moths occur from Aug. to Oct. Eggs overwinter, hatching late in spring. Larva feeds mostly on shrubs and small trees. Pupae form on leaves in loose cocoons.

1.1-1.8"

1.5-2.0"

larva

larva

LARGE MAPLE SPANWORM prefers maple and oak to its many other food plants. The larva resembles a twig. There are two broods. Moths occur through the summer. Pupa overwinters.

0.8-0.9"

THREE-SPOTTED FILLIP is single-brooded. Moths occur in early summer. Larva is green with narrow, broken, white lines and a yellow stripe on each side of the back. It feeds on maple.

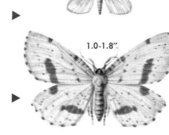

1.0-1.8"

CROCUS GEOMETER, or Cranberry Looper, feeds on many low-growing plants. Moths occur in early summer. They vary greatly in amount of wing spot- ting. Females are often spotless. Pupa overwinters.

143

0.8-1.1"

larva feeding

pupal case

BAGWORM, or Evergreen Bagworm, is common from Mass. and Kans. south to Fla. and Tex. Feeds mainly on evergreens as well as locust, sycamore, and willow. Eggs overwinter in bag. Moths emerge in fall.

BAGWORM MOTHS are named for the silken bag which the larva covers with bits of the food plant. When full-grown, it fastens the bag to a twig and changes to a pupa. The female moth is wingless and legless. She lays her eggs inside the bag. There are about 20 species of bagworm moths in North America.

CLEARWING MOTHS, with transparent wings, resemble bees and wasps. They fly in the daytime and feed at flowers. Larvae are borers in bark, stems, or roots of trees, or in stems or roots of smaller plants. A few of the 125 North American species are pests.

PEACH TREE BORER attacks stone fruit trees near the ground. It occurs in southern Canada and the United States. Color pattern varies throughout its range. It is single-brooded.

SQUASH VINE BORER occurs in most of North America except along the Pacific coast. Larva feeds in squash stems and is a serious pest. It overwinters in the ground; pupates in the spring.

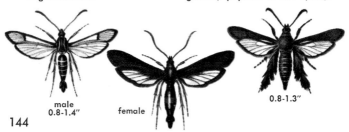

male
0.8-1.4"

female

0.8-1.3"

0.8-1.5"   0.6-1.0"

SADDLEBACK CATERPIL-LAR, named for the oval brown spot on its back, has distinctive stinging spines. It feeds on various plants, including corn, rose, cherry, and Pawpaw.

SPINY OAK-SLUG also has stinging spines. Besides oak, it feeds on pear, willow, cherry, and other trees. The moth occurs in June. The number of green spots on the forewing varies.

SLUG CATERPILLAR MOTHS number over 40 North American species. The larvae crawl like slugs; their thoracic legs are small and instead of prolegs they have sucking discs. The oval to spherical cocoon, made of dark brown silk, has a lid at one end which the emerging moth pushes aside. Larvae overwinter in the cocoons.

HAG MOTH feeds mostly on shrubs. The larva, sometimes called Monkey Slug, has projections bearing stinging hairs along its sides. These hairs are woven into the cocoon.

SKIFF MOTH occurs in midsummer. Eggs are flat and waferlike. Larva, without hair or stinging spines, prefers oak, Wild Cherry, and sycamore. Has distinct races varying in color.

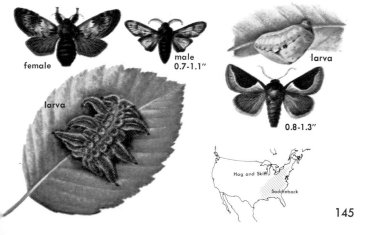

female

male
0.7-1.1"

larva

larva

0.8-1.3"

Hag and Skiff

Saddleback

145

Ragweed
Plume Moth
0.4-0.5"

lid

cocoon

larva

Puss Caterpillar
0.9-1.8"

PLUME MOTHS are named for the plume-like divisions of their wings. The forewing may be separated into two parts, the hindwing into three. Ragweed Plume Moth occurs widely in the U.S. Larva and pupa are hairy. There are over 100 species in North America.

FLANNEL MOTHS are named for the texture of the wings. The larva, about an inch long when full-grown, is slug-like and bears stinging hairs. It has seven pairs of prolegs. The Puss Caterpillar of eastern U.S. is white when young. It overwinters in the cocoon.

LEAF ROLLERS make up a very large family, including many serious pests. The larva rolls a leaf or leaves together and lives inside them. Cocoons are made of thin, soft silk on or near the food plant.

SPRUCE BUDWORM, one of our worst forest pests, feeds mainly on firs and spruces. It occurs in Canada, northern U.S., and Colorado. Eggs are laid in midsummer. Larvae overwinter.

FRUIT TREE LEAF ROLLER occurs in apple-growing areas of the U.S. and Canada. Eggs are laid in masses on tree limbs, where they overwinter. They hatch early in spring.

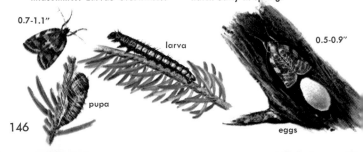

0.7-1.1"

larva

pupa

0.5-0.9"

eggs

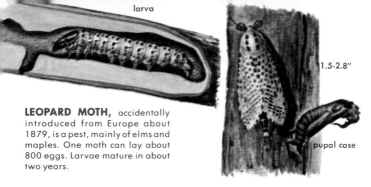

larva

1.5-2.8"

pupal case

LEOPARD MOTH, accidentally introduced from Europe about 1879, is a pest, mainly of elms and maples. One moth can lay about 800 eggs. Larvae mature in about two years.

CARPENTERWORMS are grub-like larvae that bore in trees, even in solid wood. They pupate within the bored tunnels. When emerging, moths, which look like Sphinx Moths (p.82), shed the pupal skin at the tunnel exit. They lay eggs on bark or in tunnels from which they came. Some 40 species occur north of Mexico.

CARPENTERWORM MOTH is a pest of many hardwoods, mainly ash, oak, elm, locust, and maple. Eggs are laid on bark, often near wounds. This moth has a life cycle of from three to four years.

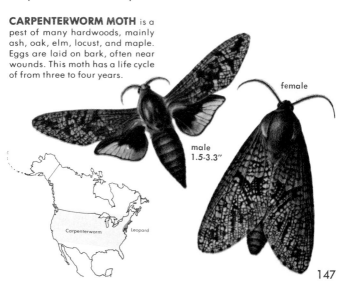

female

male
1.5-3.3"

Carpenterworm   Leopard

147

0.6-1.2"

0.8-1.3"

GRAPE LEAF FOLDER is common from Mexico into southern Canada. Female has two white spots on hindwing. It is double-brooded. Pupa overwinters in folded grape or woodbine leaves.

ZIMMERMAN PINE MOTH occurs in no. U.S. and so. Canada, attacking pines. Larvae often feed and pupate on twigs infested by another insect. Eggs or larvae overwinter.

SNOUT MOTHS make up a large family of nearly 1,000 species north of Mexico. Mostly medium- to small-sized moths, they vary greatly in appearance. The main common feature is the snout-like projection in front of the head, composed of mouthparts called palpi. Many snout moths are important pests of crops and stored grain. The larvae usually live in the stems or rolled leaves of the food plant. Most kinds make thin cocoons.

EUROPEAN CORN BORER occurs in corn-growing areas of North America. Larva bores into stalks and ears of corn. It also feeds on about 200 other plants, including dock, millet, pigweed, sorghum, and dahlia. Maturing in about a month, the larva overwinters in its tunnel and pupates there in spring.

male

female

0.8-1.4"

larva

pupa

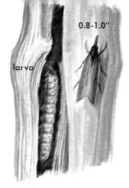

0.8-1.0"

larva

0.8-1.5"

larva

cocoon

SOUTHERN CORNSTALK BORER occurs from Md. and Kans. south to Fla. and Mex. and west to Ariz. Larva hibernates in corn stalks just above the roots. The species is double-brooded.

GREATER WAX MOTH is a nation-wide pest of beehives. Eggs are laid on or near the comb, on which the larva feeds. There are 3 broods. Winter is passed as a pupa.

CELERY LEAF TIER is widely distributed. It survives cold winters only in greenhouses. Besides celery, it is a pest of many garden and greenhouse plants, feeding and pupating in leaves.

GARDEN WEBWORM, a pest of alfalfa and other crops, ranges from S. America to so. Canada. Webbing is conspicuous in badly infested fields. Several broods occur. Pupa passes winter in soil.

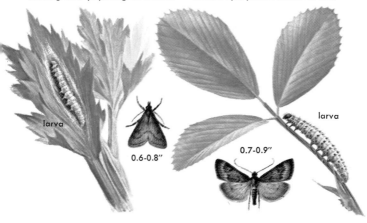

larva

0.6-0.8"

larva

0.7-0.9"

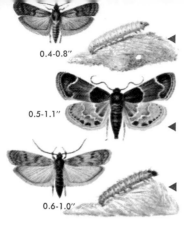

INDIAN-MEAL MOTH feeds on many kinds of stored foods. The larva webs the materials together, making a thin cocoon. Indoors it breeds continuously.

0.4-0.8"

MEAL MOTH is common and widespread. The dirty-white larva lives in a silken tube. It feeds most commonly on cereals and cereal products.

0.5-1.1"

MEDITERRANEAN FLOUR MOTH feeds on whole grains and cereals but prefers flour, which the pinkish larvae web around them in masses. Since 1890 it has spread across all of North America.

0.6-1.0"

CASE BEARER larva lives in a thin, tough sac. When full-grown, it uses the case as a cocoon. Of more than 100 species, two of the most common, the Cigar and Pistol Case Bearers, are pests of apple and other fruit trees.

LEAF MINERS damage leaves. The tiny, flat larvae feed on the inner tissue. Most of the more than 200 North American species have shining scales and plumes. One of the best known is the Solitary Oak Leaf Miner.

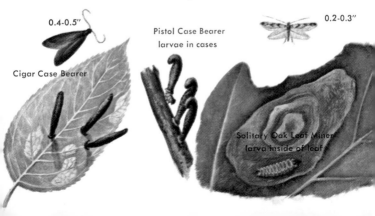

0.4-0.5"

Cigar Case Bearer

Pistol Case Bearer larvae in cases

0.2-0.3"

Solitary Oak Leaf Miner larva inside of leaf

0.6-0.8"

0.3-0.5"

larva

larva in seed

STRAWBERRY LEAF ROLLER occurs from the Atlantic west to Colo., Idaho, and so. Canada. It hibernates as a larva in a cocoon. The larva feeds in folded leaves. There are 2 to 3 generations.

MEXICAN JUMPING BEANS are seeds, usually of a species of croton from Mexico, that contain an active larva of a relative of the Codling Moth. Larvae overwinter in the seeds.

OLETHREUTID MOTHS closely resemble Leaf Rollers (p. 146) and Gelechiids (p. 152). There are over 700 North American species, which may differ greatly in their habits. Many are serious pests of farm and forest.

ORIENTAL FRUIT MOTH, of the East and Southwest, chiefly attacks peaches. The larva bores directly into the twigs early in the spring; later generations enter the fruit. Larva winters in a thin cocoon.

CODLING MOTH occurs where apples grow. After wintering as larvae in cocoons, moths emerge in early spring. The larvae infest pears and other fruits of the apple family, also English Walnuts.

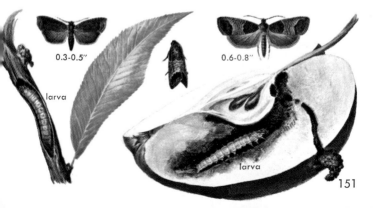

0.3-0.5"

larva

0.6-0.8"

larva

bud injury

pitch mass

0.5-0.8"

0.5-0.8"

EUROPEAN PINE SHOOT MOTH,

a serious pest of Red, Scotch, Austrian, and Mugho Pines, lays eggs in early summer near the twig tips. Larva feeds in needles at first, later enters buds. Pitch forms over their burrows. Larva overwinters in buds.

PITCH TWIGMOTH

emerges in late May and June. Eggs are laid on the twigs of Virginia and other hard pines to which the larva bores. Pitch masses form where the larva tunnels. The partly grown larva overwinters in twigs and feeds in spring.

GELECHIID MOTHS

(below) number more than 400 species in North America. Many of these small moths are pests. Some larvae feed in stems. Others feed in rolled leaves or in leaves tied together. A few mine needles. Forewings are narrow, hindwings somewhat angular.

GOLDENROD SPINDLE GALL

might be confused with the round gall of a fly. Before pupating in the gall, the larva makes an exit hole. Moths emerge through it in the fall.

ANGOUMOIS GRAIN MOTH

is a serious pest of stored, whole-kernel grains, especially wheat and corn, throughout the United States. Some grain is attacked in the field at harvest time.

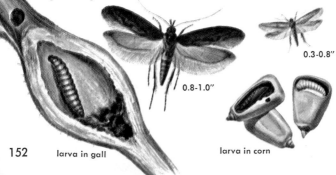

0.8-1.0"

0.3-0.8"

152   larva in gall

larva in corn

Webbing Clothes Moth
0.3-0.6"

Carpet Moth
0.3-0.6"

Case-making
Clothes Moth

0.3-0.6"

larva

TINEID MOTHS include more than 125 North American species. Some eat leaves, some fungus, and some wool fabric or fur. The Carpet Moths and Clothes Moths belong to this family. The Webbing Clothes Moth is most common. The Case-making Clothes Moth is darker. Its larva lives in a movable, open-ended case.

OTHER MOTHS   The moths and butterflies on the previous pages represent thirty-six families of Lepidoptera. Some thirty more families exist in North America, including many inconspicuous moths. The following two moths are exceptions.

MIMOSA WEBWORM occurs on mimosa and Honey Locust in the East. It ties leaves together and feeds within the webbing. Double-brooded, it overwinters as a pupa in a soft, white cocoon.

YUCCA MOTH is essential for the Yucca's reproduction. The female moth carries pollen to the stigma, fertilizing the eggs which form seeds. Larva feeds in the seeds, pupates in spring.

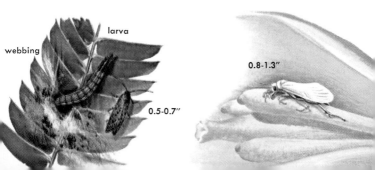

webbing

larva

0.5-0.7"

0.8-1.3"

SCIENTIFIC NAMES

19 Red-spotted Purple:
   Limenitis arthemis astyanax
21 Pipevine: *Battus philenor*
   Polydamas: *B. polydamas*
22 Black: *Papilio polyxenes*
   Baird's: *P. bairdii*
   Anise: *P. zelicaon*
23 Indra: *P. indra*
24 Palamedes: *P. palamedes*
   Alaskan: *P. machaon aliaska*
25 *P. cresphontes*
26 *P. troilus*
27 Eastern: *P. glaucus*
   Western: *P. rutulus*
28 Two-tailed: *P. multicaudata*
   Pale: *P. eurymedon*
   Zebra: *Eurytides marcellus*
29 Clodius: *Parnassius clodius*
   Smintheus: *P. phoebus smintheus*
30 Colias eurytheme
31 Clouded: *C. philodice*
   Pink: *C. interior*
32 Sara: *Anthocharis sara*
   Falcate: *A. midea*
   Olympia: *Euchloe olympia*
33 Orange: *Phoebis philea*
   Cloudless: *P. sennae*
34 Calif.: *Zerene eurydice*
   Southern: *Z. cesonia*
   Little: *Eurema lisa*
35 Sleepy: *E. nicippe*
   Fairy: *E. daira*
   Dainty: *Nathalis iole*
   Mustard: *Pieris napi*
36 Pine: *Neophasia menapia*
   Great: *Ascia monuste phileta*
   Giant: *Ganyra josephina*
   Florida: *Appias drusilla neumoegenii*
37 Checkered: *Pieris protodice*
   Cabbage: *P. rapae*
38 *Danaus plexippus*
39 *D. gilippus berenice*
40 Eyed: *Lethe eurydice*
   Creole: *L. creola*
   Pearly: *L. portlandia*
41 Little: *Megisto cymela*
   Gemmed: *Cyllopsis gemma*
   Carolina: *Hermeuptychia hermes*
   Georgia: *Neonympha areolatus*
   Plain: *Coenonympha inornata*
   Calif.: *C. california*
42 Riding's: *Neominois ridingsii*
   Common: *Erebia epipsodea*
   Common Wood: *Cercyonis pegala*
   Nevada: *Oeneis nevadensis*
43 Gulf: *Agraulis vanillae*
   Julia: *Dryas julia*
   Zebra: *Heliconius charitonius*
44 Variegated: *Euptoieta claudia*
   Regal: *Speyeria idalia*
45 Great: *S. cybele*
   Aphrodite: *S. aphrodite*
46 Diana: *S. diana*
   Nevada: *S. callippe nevadensis*

Eurynome: *S. mormonia eurynome*
47 Silver: *Boloria selene myrina*
   Eastern: *B. bellona toddi*
   Western: *B. epithore*
48 Baltimore: *Euphydryas phaeton*
   Silvery: *Charidryas nycteis*
   Harris': *C. harrisii*
   Chalcedon: *Occidryas chalcedona*
49 Mylitta: *Phyciodes mylitta*
   Field: *P. campestris*
   Phaon: *P. phaon*
   Pearl: *P. tharos*
50 Question Mark: *Polygonia interrogationis*
   Satyr: *P. satyrus*
51 Comma: *P. comma*
   Fawn: *P. faunus*
   Zephyr: *P. zephyrus*
52 *Junonia coenia*
53 Painted: *Vanessa cardui*
   West Coast: *V. carye*
54 Red Admiral: *V. atalanta*
   American: *V. virginiensis*
55 Compton: *Nymphalis j-album*
   Milbert's: *N. milberti*
   Mourning: *N. antiopa*
56 White: *Limenitis arthemis arthemis*
   Viceroy: *L. archippus*
57 Weidemeyer's: *L. weidemeyerii*
   Lorquin's: *L. lorquini*
   Red-spotted: *L. arthemis astyanax*
58 Calif.: *Adelpha bredowii*
   Goatweed: *Anaea andria*
   Morrison's: *A. aidea morrisonii*
59 Hackberry: *Asterocampa celtis*
   Tawny: *A. clyton*
60 Florida: *Eunica tatilista tatilista*
   Ruddy: *Marpesia petreus*
   Mimic: *Hypolimnas misippus*
61 Little: *Liephelisca virginiensis*
   Swamp: *L. muticum*
   Northern: *L. borealis*
62 Nais: *Apodemia nais*
   Mormon Dark: *A. mormo mormo*
   Mormon Light: *A. mormo mejicanus*
   Snout: *Libytheana bachmanii*
63 Gray: *Strymon melinus*
   Great: *Atlides halesus*
   Colorado: *Hypaurotis chrysalus*
64 White-M: *Parrhasius m-album*
   Edward's: *Satyrium edwardsii*
   Acadian: *S. acadica*
   Southern: *S. favonius*
   Red-banded: *Calycopis cecrops*
   Coral: *Harkenclenus titus*
65 Calif.: *Satyrium californicum*
   Banded: *S. calanus falacer*
   Hedgerow: *S. saepium*
   Striped: *S. liparops*
   Sylvan: *S. silvinum*
   Olive: *Mitoura gryneus*
66 West: *Incisalia eryphon*
   Banded: *I. niphon*
   Brown: *I. augustus*

67 Western: *I. iroides*
Hoary: *I. polios*
Henry's: *I. henrici*
Frosted: *I. irus*
68 American: *Lycaena hypophlaeas*
Great: *L. xanthoides*
Ruddy: *L. rubidus*
69 Purplish: *L. helloides*
Gorgon: *L. gorgon*
Bronze: *L. thoe*
70 Harvester: *Feniseca tarquinius*
Eastern: *Everes comyntas*
71 Western: *E. amyntula*
Pygmy: *Brephidium exilis*
Dwarf: *B. isophthalma*
Common: *Celastrina ladon*
72 Marine: *Leptotes marina*
Reakirt's: *Hemiargus isola*
Acmon: *Plebejus acmon*
Orange: *P. melissa*
73 Saepiolus: *P. saepiolus*
Silvery: *Glaucopsyche lygdamus*
Sonora: *Philotes sonorensis*
Square-spotted: *P. battoides*
74 Hoary: *Achalarus lyciades*
Silver: *Epargyreus clarus*
75 Northern: *Thorybes plyades*
Southern: *T. bathyllus*
Golden: *Autochton cellus*
Long-tailed: *Urbanus proteus*
76 Sleepy: *Erynnis brizo*
Dreamy: *E. icelus*
Juvenal's: *E. juvenalis*
Mottled: *E. martialis*
Mournful: *E. tristis*
Funereal: *E. funeralis*
77 Checkered: *Pyrgus communis*
Grizzled: *P. centaureae*
Braz.: *Calpodes ethlius*
Common: *Pholisora catullus*
Southern: *Staphylus hayhurstii*
78 Least: *Ancyloxypha numitor*
Uncas: *Hesperia uncas*
Cobweb: *H. metea*
Juba: *H. juba*
Indian: *H. sassacus*
Leonard's: *H. leonardus*
Golden: *Copaeodes aurantiaca*
79 Broken Dash: *Wallengrenia otho*
Long Dash: *Polites mystic*
Vernal: *Pompeius verna*
Peck's: *Polites coras*
Fiery: *Hylephila phyleus*
Field: *Atalopedes campestris*
80 Hobomok: *Poanes hobomok*
Zabulon: *P. zabulon*
Roadside: *Amblyscirtes vialis*
Ocola: *Panoquina ocola*
Yucca: *Megathymus yuccae*
81 Paratrea plebeja
82 Pink: *Agrius cingulatus*
Carolina: *Manduca sexta*
83 Rustic: *Manduca rustica*
Five-spotted: *M. quinquemaculata*

84 Waved: *Ceratomia undulosa*
Catalpa: *C. catalpae*
85 Four-horned: *C. amyntor*
Hermit: *Sphinx eremitus*
Pawpaw: *Dolba hyloeus*
86 Elegant: *Sphinx perelegans*
Great Ash: *S. chersis*
87 Laurel: *S. kalmiae*
Apple: *S. gordius*
Wild Cherry: *S. drupiferarum*
88 Ello: *Erinnyis ello*
Abbot's: *Lapara coniferarum*
Northern: *L. bombycoides*
89 Big Poplar: *Pachysphinx modesta*
Abbot's: *Schecodina abbotti*
Sequoia: *Sphinx sequoiae*
90 One-eyed: *Smerinthus cerisyi*
Huckleberry: *Paonias astylus*
Small-eyed: *P. myops*
91 Blinded: *P. excaecatus*
Twin-spot: *Smerinthus jamaicensis*
Walnut: *Laothoe juglandis*
92 Nessus: *Amphion floridensis*
Hydrangea: *Darapsa versicolor*
Azalea: *D. pholus*
Hog: *D. myron*
93 Achemon: *Eumorpha achemon*
Pandorus: *E. satellitia*
Lesser: *E. fasciata*
94 Hummingbird: *Hemaris thysbe*
Snowberry: *H. diffinis*
White-lined: *Hyles lineata*
Galium: *H. gallii*
95 *Samia cynthia*
96 *Hyalophora cecropia*
97 Glover's: *H. gloveri*
Ceanothus: *H. euryalus*
98 Columbia: *H. columbia*
Pandora: *Coloradia pandora*
99 Antheraea polyphemus
100 *Callosamia promethea*
101 *C. angulifera*
102 *Actias luna*
103 Io: *Automeris io*
Sheep: *Hemileuca eglanterina*
104 Buck: *H. maia*
Nevada: *H. nevadensis*
105 Range: *H. oliviae*
Silk: *Bombyx mori*
106 Honey: *Sphingicampa bicolor*
Rosy: *Dryocampa rubicunda*
107 Orange: *Anisota senatoria*
Spiny: *A. stigma*
Pink: *A. virginiensis*
108 *Citheronia regalis*
109 *Eacles imperialis*
110 *Ecpantheria scribonia*
111 Acrea: *Estigmene acrea*
Isabella: *Pyrrharctia isabella*
112 Garden: *Arctia caja*
Fall: *Hyphantria cunea*
113 Clymene: *Haploa clymene*
Arge: *Apantesis arge*
Virgo: *A. virgo*

114 Hickory: *Lophocampa caryae*
Spotted: *L. maculata*
Pale: *Halisidota tessellaris*
115 Yellow: *Spilosoma virginica*
Ranchman's: *Platyprepia guttata*
Dogbane: *Cycnia tenera*
116 Showy: *Holomelina ostenta*
Milkweed: *Euchaetes egle*
Bella: *Utetheisa bella*
117 Brown: *Ctenucha brunnea*
Virginia: *C. virginica*
Eight-spotted: *Alypia octomaculata*
Calif.: *Phryganidia californica*
118 *Ascalapha odorata*
119 American: *Acronicta americana*
Smeared: *A. oblinita*
Cottonwood: *A. lepusculina*
120 Black: *Agrotis ipsilon*
Pale-sided: *A. malefida*
Spotted: *Xestia dolosa*
W-marked: *Spaelotis clandestina*
121 Pale: *Agrotis orthogonia*
Spotted-sided: *Xestia badinodis*
Glassy: *Crymodes devastator*
Striped: *Lacanobia legitima*
122 Dark-sided: *Euxoa mesoria*
Striped: *E. tessellata*
Bronzed: *Prorella emmedonia*
White: *Pleonectapoda scandens*
123 Dingy: *Feltia subgothica*
Bristly: *Lacinipolia renigera*
Variegated: *Peridroma saucia*
Yellow: *Spodoptera ornithogalli*
124 Fall: *S. frugiperda*
Armyworm: *Pseudaletia unipuncta*
Beet: *Spodoptera exigua*
Wheat: *Thurberiphaga diffusa*
125 Army: *Chorizagrotis auxiliaris*
Zebra: *Melanchra picta*
Cotton: *Alabama argillacea*
Green: *Plathypena scabra*
126 Alfalfa: *Autographa californica*
Bilobed: *A. biloba*
Cabbage: *Trichoplusia ni*
Celery: *Anagrapha falcifera*
127 Forage: *Caenurgina erechtea*
Lunate: *Zale lunata*
Green: *Lithophane antennata*
Dried: *Idia lubricalis*
128 Widow: *Catocala vidua*
White: *C. relicta*
129 Aholibah: *C. aholibah*
Tiny: *C. micronympha*
Penitent: *C. piatrix*
130 Locust: *Euparthenos nubilis*
Copper: *Amphipyra pyramidoides*
Pearly: *Eudryas unio*
131 White-veined: *Simyra henrici*
Stalk: *Papaipema nebris*
Corn: *Heliothis zea*
132 White-marked: *Clostera albosigma*
Poplar: *C. inclusa*
Tentacled: *Cerura scitiscripta multiscripta*

133 Yellow: *Datana ministra*
Walnut: *D. integerrima*
Sumac: *D. perspicua*
134 Saddled: *Heterocampa guttivitta*
Variable: *Lochmaeus manteo*
Anguina: *Dasylophia anguina*
Elm Leaf: *Nerice bidentata*
135 Red-humped Caterpillar: *Schizura concinna*
Unicorn: *S. unicornis*
Rough: *Nadata gibbosa*
Red-humped Oakworm: *Symmerista albifrons*
136 Satin: *Leucoma salicis*
Gypsy: *Lymantria dispar*
137 White-marked: *Orgyia leucostigma*
Western: *O. vetusta*
Rusty: *O. antiqua*
Pine: *Dasychira plagiata*
138 Eastern: *Malacosoma americanum*
Western: *M. californicum*
139 Forest: *M. disstria*
Wild Cherry: *Apatelodes torrefacta*
140 Fall: *Alsophila pometaria*
Spring: *Paleacrita vernata*
Bruce: *Operophtera bruceata*
141 Spear-marked: *Rheumaptera hastata*
Cherry: *Hydria undulata*
Currant: *Itame ribearia*
Walnut: *Phigalia plumogeraria*
142 Linden: *Erannis tiliaria*
Pepper-and-salt: *Biston cognataria*
Elm: *Ennomos subsignarius*
Hemlock: *Lambdina fiscellaria*
143 Chain-spotted: *Cingilia catenaria*
Large Maple: *Prochoerodes transversata*
Three-spotted: *Heterophleps triguttaria*
Crocus: *Xanthotype sospeta*
144 Bagworm: *Thyridopteryx ephemeraeformis*
Peach Tree: *Sanninoidea exitiosa*
Squash: *Melittia curcurbitae*
145 Saddleback: *Sibine stimulea*
Spiny: *Euclea delphinii*
Hag: *Phobetron pithecium*
Skiff: *Prolimacodes badia*
146 Ragweed: *Adaina ambrosiae*
Puss: *Megalopyga opercularis*
Spruce: *Choristoneura fumiferana*
Fruit Tree: *Archips argyrospilus*
147 Leopard: *Zeuzera pyrina*
Carpenterworm: *Prionoxystus robiniae*
148 Grape Leaf: *Desmia funeralis*
Zimmerman: *Dioryctria zimmermani*
European: *Ostrinia nubilalis*
149 Southern: *Diatraea crambidoides*
Greater: *Galleria mellonella*
Celery: *Udea rubigalis*
Garden: *Achyra rantalis*

INDEX

Asterisks (*) denote pages on which illustrations appear.

MEASURING SCALE (IN MILLIMETERS AND CENTIMETERS)

T